Environmental Agencies
in the United States

Environmental Agencies in the United States

The Enduring Power of Organizational Design and State Politics

By JoyAnna Hopper

LEXINGTON BOOKS
Lanham • Boulder • New York • London

Published by Lexington Books
An imprint of The Rowman & Littlefield Publishing Group, Inc.
4501 Forbes Boulevard, Suite 200, Lanham, Maryland 20706
www.rowman.com

6 Tinworth Street, London SE11 5AL, United Kingdom

British Library Cataloguing in Publication Information Available

Library of Congress Cataloging-in-Publication Data
Names: Hopper, JoyAnna S., 1989- author.
Title: Environmental agencies in the United States : the enduring power of
    organizational design and state politics / by JoyAnna Hopper.
Description: Lanham : Lexington Books, [2020] | Includes bibliographical references
    and index. | Summary: "In Environmental Agencies in the U.S., the author considers
    mandates assigned to environmental agencies and how those mandates shape
    environmental enforcement. Arguing the importance of structure, organizational
    norms, and state politics, the author crafts a nuanced explanation of the policy
    differences that shape Americans' well-being"—Provided by publisher.
Identifiers: LCCN 2020001356 (print) | LCCN 2020001357 (ebook) | ISBN
    9781498573474 (cloth) | ISBN 9781498573498 (pbk) | ISBN 9781498573481 (epub)
Subjects: LCSH: Environmental agencies—United States—States. | Environmental
    protection—Management. | Public administration—United States—States. |
    Environmental policy—United States—States.
Classification: LCC TD171 .H66 2020 (print) | LCC TD171 (ebook) | DDC 333.70973—
    dc23
LC record available at https://lccn.loc.gov/2020001356
LC ebook record available at https://lccn.loc.gov/2020001357

This book is dedicated to my mother and father,
Katie and Jimmy Hopper and to my brother,
Kelly Hopper. Thank you for all of your love and support.

Contents

List of Tables and Figures

Preface

On a morning in mid-February, I sat across a coffee house table from a woman who appeared exasperated. Her exasperation was in response to a question I had posed to her about her job working in a state-level environmental protection agency. I had asked her the question, "Do you feel as if the work you perform is supported by your colleagues in the department." Her response, after a stifled chuckle was, "which kind?" Confused by her response, I pushed her further, asking what she meant. Did she mean "which kind of work," "which kind of support?" She stopped me short, stating clearly, "I mean—which kind of *colleagues*." She went on to explain that in her agency there were *health* people and there were *regulation* people, and that the two groups of individuals rarely agreed on much. I asked her why these groups felt as if they were pursuing competing goals; she shrugged and offered, "they just speak different languages."

This statement—that sub-groups within environmental agencies simply "don't speak the same language"—would be a statement I heard often during my interviews with state-level agency workers between 2014–2019. And, I am not alone. In Norton's (2005) interdisciplinary evaluation of sustainability theory, he notes that Environmental Protection Agency (EPA) employees "from different backgrounds and disciplines continued to interact, carry on conversations, and do their jobs, hardly noticing that they spoke languages without available translations. They, like unattended visitors at the EPA building, wandered into blind corridors and, when they asked how one might get to a more rational environmental policy, were too often told, 'You can't get there from here'" (Norton 2005). This difficulty in translation reflects a historical set of fractures within environmental protection in the United States—a fracturing that reveals itself often when policy problems necessitate that we decide what the goals of protecting the environment should be and how those goals might be obtained. In her seminal multi-edition text, *The Environmental Case*, Judith Layzer states that all debates surrounding environmental policy are debates about values—and that it is those debates over values that have created such contention around environmental policy making and implementation. Layzer is clear that the "black and white" understanding of liberals being pro-environment and conservatives being pro-industry, in no way truly represents the dimension of environmental opinions that inform and pressure policy makers in the United States. Preservationism, conservationism, concerns for public health, concerns for the economy, concerns over justice, and debates over appropriate regulation color the debates over environmental policy in the United States, representing the conflict of values that Layzer describes. Additionally, the intentional fracturing of environmental policy making and implementation introduced by environmental federalism allows this diversity of values related to environmental protection to flourish, leading to sometimes drastically different approaches to protecting the environment. The day-to-day decisions that determine whether Americans' air and water is clean are made by individuals whose understanding of and preferences for regulatory processes differ by the state, by the agency, and by the department within the agency.

These differences have meaningful implications. In 2013, MIT researchers found that ground-level emissions from smokestacks, marine and rail operations, automobiles, and commercial/residential heating lead to around 200,000 early deaths in the United States every year (Caiazzo et al. 2013). According to these researchers, individuals dying early from air pollution die about a decade earlier than they would without exposure. They also found that these deaths vary by location in the United States. Yes, the urban and rural divide is relevant, but region matters too. What kind of industries are most prevalent? How does city planning position roadways and align commercial and residential areas? How do historical and current patterns of segregation lead to particular segments of the population receiving higher levels and more dangerous types of exposure? And, in what ways does public policy address or fail to address deadly pollution? These factors do not only differ from city-to-city, but also from state to state, where dependence on natural resources and manufacturing, topography, population density, and politics differ significantly.

For some time, academics and popular media have pointed to political control by Republicans or Democrats as the primary motivating factor driving pollution control policy in the United States. Liberal states like California are leaders in environmental policy, while conservative states like Texas are laggards (Lester 1995). But, this distinction is somewhat misleading. For example, according to the MIT research I referenced previously, California suffers the worst health impacts of air pollution (Chu 2013). And, Texas—as I note in a later chapter—boasts one of the most successful water conservation programs in the nation. The determinants of environmental policy and the effectiveness of those policies are numerous and complex. Part of that complexity originates in the fact that many environmental policy decisions are not made by elected officials. While the liberal-conservative distinction may help to explain legislation, executive orders, and court decisions, it cannot fully explain how those policies are implemented. Environmental agencies and offices across the United States are delegated with the authority to create the programs and rules necessary to see through environmental policy decisions made by political actors. And, these bureaucratic actors do not find themselves working within indistinguishable organizations. The structures, norms, and values of environmental agencies differ across the American states, and these differences are meaningful determinants of environmental policy behavior because of the powerful impact that organizational characteristics can have on those individuals who work within them. If we continue to discount the complexity that these differences introduce into our understanding of environmental policy in the United States, we will be unable to undertake the reforms and policy changes necessary to mitigate environmental tragedy.

In this book, I seek to evaluate what drives the actions of environmental protection agencies in the United States. Particularly, I focus on the ways in which conflicts surrounding the value systems inherent to the ideologies of conservation, public health, and environmental regulation have led to a diversity of approaches in protecting the environment. I argue that the structures and values of American state-level environmental agencies—structures and values that nurture particular perspectives on environment and regulation—affect environmental agency employees' efforts at collaboration, the way they seek common ground with industry, and the decisions they make about when and how to issue punishment for the violation of environmental rules. Structures and values are not short-term organizational forces; they perpetuate across time, continuing to affect "the way things are done around here" (Wilson 1989) as each year of environmental enforcement activity informs the next. While political scientists and students of public administration have often cited political principals as powerful motivators of environmental enforcement behavior, political principals are only as powerful as their ability to penetrate the long-standing structures and cultures of the bureaucratic agencies they seek to influence. The preferences of state legislators, governors, and even the federal Environmental Protection Agency must filter through the rules, norms, and values of the state-level employees who perform inspections, uncover violations, and negotiate enforcement.

In the introductory chapter, I discuss a short history of environmental protection in the American states and the development of environmental protection agencies, followed by my proposal of a theory of environmental agency design in chapter 2. The second chapter illustrates why organizational choices often insulate and reinforce particular ideologies and how organizational choices that combine tangentially related goals can lead to the emphasis of one set of goals over another. I analyze the role of politics in organizational choices and in motivating agency behavior, emphasizing the difficulty political principals may have in shifting the long-standing organizational norms and values of agents. These conversations are accompanied by environmental enforcement data that reinforces the complexity and diversity of environmental decision making in the American states.

In the third, fourth, and fifth chapters I describe each of the existing environmental agency types: the natural resource conservation and environmental protection agency (NREP), the public health and environmental protection agency (PHEP), and the mini-EPA. Using case studies from across the states and interviews with state agency workers, I describe how differences in agency design/organization affect the views and behavior of environmental agency workers.

In the sixth, seventh, and eighth chapters, I further test my hypotheses regarding the effects of environmental agency design on enforcement. In chapter 6, I perform content analyses on environmental agency documents to determine what agency mandates are emphasized in combined environmental agencies. The measure I construct indicates that agency design is related to the way agencies speak about their day-to-day activities and what activities they publicize and prioritize. In chapter 7, I evaluate whether that organizational design and its accompanying organizational values affect day-to-day enforcement activities, such as the number of enforcement actions taken and the severity of monetary penalties. I find that agency design choices, other than the mini-EPA design, lead to more collaborative and cooperative approaches with industry—often meaning fewer and more flexible enforcement actions. In chapter 8, I evaluate how the power or capacity of environmental agencies helps or hinders agencies' ability to challenge the preferences of political principals who are aimed at constraining enforcement. I find that a more powerful and capable agency is able to translate their preferences for enforcement, even in the face of a powerful state legislature. I close the book by speaking about the future of environmental protection in the United States and what the EPA's movement toward management-based regulation may mean for environmental state agencies.

As Wilson (1989) expresses, I wish that this book presented a "simple, elegant, comprehensive theory" of environmental agency behavior in the United

States, but my findings seem to complicate the picture more than simplify it. That said, I believe the acknowledgment of that complexity is potentially the book's greatest contribution. Scholars of environmental policy, such as myself, have often jumped too quickly from political control to environmental effects, without acknowledging the numerous bureaucratic characteristics that may influence the implementation of environmental policy. This book is simply another step forward in addressing that missing component.

I have many to thank for their assistance and support as I was writing this book. First and foremost, I would like to thank the environmental agency employees who took the time to speak with me and to allow me to include their insights as part of my research. Without them, this effort would have been impossible. Importantly, to protect those contributors, interviewees throughout the book are completely anonymized. I also must thank those who supported this research from the beginning, including Dr. Lael Keiser, Dr. Peverill Squire, and other mentors and friends at the University of Missouri. Additionally, I am forever grateful to my husband, Clint, our two dogs, Lady and Daisy, and our two cats, Josie and Zeva, for patience, love, and understanding as I have worked long hours to finish this project. Lastly, many scholars have contributed their knowledge and expressed confidence in my work; I owe much to the political science community. I especially want to thank those that are part of the State Politics and Policy section of the American Political Science Association and the reviewers of this book and related projects, from which I have always received constructive and formative feedback.

Chapter One

Environmental Agencies in the United States

State and local governments played the role of primary regulator of air and water pollution between the first half of the nineteenth century and the creation of the EPA in 1970. Some water pollution control policies even date back to the colonial period as cities sought to control the disposal of sewage (Davies 1970). Once the connection between water pollution and contagious disease was established, states became more active, creating boards of health and attempting to partner with localities to control pollution. These partnerships were not always fruitful, as cities did not care to control their own disposal of waste if it was simply to head downstream, and they could not control the cities up-stream that held the same incentive (Davies 1970, 121). Air pollution introduced greater complexity, as it was even less controllable than pollution moving by water. Thus, states governments began to intervene. Through public health agencies and boards, pollution control boards, and natural resource agencies and boards, states attempted to control the very industries that their economies relied upon. However, their efforts were mostly unsuccessful. Pollution control, executed via a mélange of state and local health and conservation policies, was not preventing deadly smog episodes, fiery rivers, and chemical spills. Their failures can be attributed to a lack of resources and funding (even with federal help). "Few states undertook serious pollution-control programs . . . because state and local officials were deeply concerned about fostering economic development, and because environmental activists in most states had insufficient clout to challenge economic interests" (Layzer 2015, 33). However, even as the federal government began to slowly enter the arena of environmental protection through expanded funding and advising during the mid-twentieth century, Eisenhower referred to pollution as a "uniquely local blight," claiming that the "primary responsibility for solving the problem lies not with the federal government but rather must be assumed and exercised, as it has been, by state and local governments." State governments were reluctant to relinquish power, and the federal government was reluctant to take responsibility.

This began to change in the late 1960s. Following the publication of Rachel Carson's *Silent Spring* and the highly publicized environmental disasters of deadly New York smog, the Cuyahoga river fires, and the Santa Barbara Oil Spill, the counterculture movement that had pushed for the end to the Vietnam War and an expansion of civil rights took on the cause of environmental protection. "Pollution had gotten bad enough, and science had become advanced enough to make the reasons why clear" (Rothamn 2017). Championing the environment was not a new movement, by any means; Americans had been pushing for preservation and conservation from the transcendentalists to Teddy Roosevelt. But, this building pressure brought renewed and sharpened attention to the failures of state and local environmental protection efforts. There were calls for the regulation of industry and for a larger and more aggressive federal effort. Incoming president Richard Nixon felt that pressure. In Summer 1969, Nixon created the Environmental Quality Council, which was shortly followed by Congress's Environmental Policy Act of 1969. Although Nixon attempted to address gaps in environmental protection by piecemeal executive orders, he soon realized that the number of agencies and departments that held responsibility over implementing these orders, their overlapping jurisdictions, and the inevitable turf wars made this approach unpromising. Thus, Nixon sought advice in determining an alternative.

In April 1970, Nixon received input from the Advisory Council on Executive Organization regarding the organization of anti-pollution programs. Nixon had pledged to "repair [environmental] damage already done, and to establish new criteria to guide us in the future." In response to this pledge, the council, led by Roy L. Ash, recommended to Nixon that anti-pollution programs be "merged into an Environmental Protection Administration, a new independent agency of the Executive Branch" (Ash Council Memo 1970). According to the council, this decision was made due to the inadequacy of the government structures that were currently in place, along with a growing environmental crisis that demanded a strong government response. This was not Ash's original idea. Originally, Ash had suggested that environmental protection be combined with natural resource management, as the national government's primary role up to this point had been as "conservator of wilderness" ("The Guardian: Origins of the EPA" 1992). However, after some consideration of this organizational design, Ash and the council rejected the idea of combining the regulatory actions of setting and enforcing pollution control standards with natural resource management. They also rejected the idea of combining pollution control with the Department of Health, Education, and Welfare. In arguing against these combinations—and in favor of a separate regulatory agency—the council stated that the combinations would bias the agency's setting and enforcing of pollution standards in a direction that served the goals of natural resource management or health and education (Ash Council Memo 1970). The memo refers to this as the "inherent bias" of each existing department.

Thus, when Nixon was faced with addressing the pressure to reform environmental protection by imposing and enforcing rules on polluting industries, his advisors moved away from the status quo of state and local control via natural resource boards and public health agencies. Additionally, Nixon's advisors dissuaded him from maintaining the approach of carrying out environmental protection via bureaucracies like the Public Health Service—an agency that was not singularly tasked with protecting the environment. An Environmental Protection Agency (EPA) would be created, and it would be charged with the creation and enforcement of federal environmental standards. The federal government, through an agency dedicated only to environmental protection, would hold power over enforcing environmental law. But, states and their environmental agencies had not become irrelevant. In fact, the states were to play a continued active role in implementing environmental policies.

PARTIAL PREEMPTION AND STATE INVOLVEMENT IN ENVIRONMENTAL PROTECTION POST-1970

Starting with major pieces of environmental legislation, such as the Clean Air Act (CAA) (1970), and the creation of the EPA, the federal government became an active part of regulating industrial and citizen interactions with the environment. However, federal actions then—and especially now—are not as hands-on as most might imagine. The role of the federal government in laws like the CAA is to set standards for common pollutants to protect people within an "adequate margin of safety." Their role is *not* to be the primary enforcers of those standards. Rather, states are given the opportunity to create their own implementation plans aimed at maintaining the standards set by the federal government. Authority for developing and enforcing regulatory standards is primarily given to the states; the EPA simply "sets a floor below which standards may not fall" (Konisky and Woods 2012a, 477). This is a process referred to as partial preemption. If states are unable to meet standards or do not feel equipped to implement their own plans and enforcement, they can hand that power over to the EPA. However, even with that option, according to the Environmental Council of the States (2010), states still operate about 96 percent of programs for which they have been given the option of taking primacy (Konisky and Woods 2012a).

Although primacy means potentially burdensome expenditures on environmental protection and the funneling of resources into enforcement activities that may challenge state goals of attracting industry, primacy provides states with enough control over day-to-day enforcement to execute regulation as they see fit. This is especially the case today, as more and more implementation power for environmental policies has been devolved to state agencies. In the enforcement of most federal environmental protection laws, state agencies get to determine what constitutes a violation of the law, whether a violation should be noted as severe or minor, which industries are subject to inspection and how often, and what enforcement actions are appropriate (e.g., should a letter outlining the violation and expected follow-up be sent or should a monetary penalty be assigned). Additionally, Wood (1991) finds that states may not even be all that responsive to policy interventions/pressure from the federal government. In looking at Reagan's attempts to reign in perceived overregulation, states did not seem to respond. Instead, their actions remained mostly consistent. More interestingly, though, is that it appeared that federal environmental protection actions were more motivated by state actions than the other way around. States' environmental agencies drive the direction of environmental policy. And, this is consequential because each state differs in its regulatory preferences. Each state requires a different approach to environmental protection, and each American citizen is subject to these differences.

THE DETERMINANTS OF
ENVIRONMENTAL ENFORCEMENT IN THE STATES

There is an expansive literature dedicated to explaining environmental protection decisions in the American states (Hays et al. 1996; Daley and Garand 2005; Woods 2008; Ringquist 1993; Bacot and Dawes 1997; Davis and Davis 1999; Wood 1992; Atlas 2007; Konisky and Woods 2012b; Hall and Kerr 1991; Sigman 2003; Hopper 2017; Hopper 2019; Konisky 2008; Woods 2006; Rabe 2007; Konisky 2007; Konisky 2009). Scholars of state-level environmental politics have dedicated years of work to determining how the very different state economies, political institutions, and ideologies have shaped decisions that range from the number of inspections a state performs to the environmental votes of state legislators. From this literature emerge a few main themes that we might consider in determining why environmental enforcement differs and the implications of those differences. The determinants of environmental enforcement generally fall into three categories: environmental need, state capacity, and state willingness.

To begin, states' natural environmental order is going to determine the type and extensiveness of environmental policies (and, thus, environmental enforcement) needed. As Konisky and Woods (2012a) point out, things like "air and water quality are not just a function of total emissions but also how pollution is affected by topographic features and the natural attributes of air sheds and waterways" (474). How much land and water a state has can affect the types of policies it needs *and* environmental agencies' ability to inspect and evaluate states' environmental conditions. Additionally, the way rivers flow and the way the wind blows affects how pollution spreads, also affecting the need for particular kinds of policies, programs, and enforcement activities in specific places. Minnesota's regulation of the Mississippi River at its head will be quite different from Louisiana's regulation of the River at its delta. Water use policies and enforcement differ greatly between water-rich Michigan and drought-ridden California.

In addition to the state's natural environment, however, are the environmental conditions the state may have imposed on its natural environment through industry. The structure of states' economies, whether it be driven by manufacturing, resource extraction, or service, also determines the need for particular kinds of policies and a certain level of enforcement. The more pollution produced, particularly in states that rely on commodities such as coal or have invested heavily in manufacturing, the greater the need for environmental protection actions. Thus, the scope and types of state industries are likely to drive environmental decision making. To be clear, the relationship here is not one-directional. Rather—as I describe shortly—more industry can create more pollution, requiring stricter environmental protection efforts, *or* more industry can create more economic dependency, requiring more lenient environmental protection efforts. Regardless, industry's presence has a significant effect. In sum, environmental need is driven by the state's natural environmental order, along with the relationship between industry and pollution that has altered that natural order.

In regard to *state capacity*, state agencies' ability to create and effectively implement environmental programs also drives environmental enforcement outputs. State institutions, including courts, executives, legislatures, and bureaucracies, differ significantly in regard to their professionalization, time in office/at work, susceptibility to outside pressure, powers, and the ability to wield those powers. All of these differences can affect what environmental enforcement looks like. For example, in a state where legislators are highly professionalized (well-paid, spend more time in the legislature, and have more staff support), they may be able to exert more control over environmental agencies and their interactions with industry (Squire 2017). Whereas, a weaker legislature facing a powerful governor, may have less influence over the direction of environmental agency decisions. Specifically, a state bureaucracy's ability to have some autonomy over its decision making and also bureaucracies' capacity for carrying out logistics are factors likely to have environmental enforcement implications. For example, if a state agency is perceived by legislators to be incompetent or unruly, it is likely to lose power to the legislature (Nicholson-Crotty and Miller 2012; Carpenter 2010). Thus, the variation in power, organization, norms, and rules across state institutions affects the ability of state governments to build environmental programs and implement those programs effectively and efficiently.

In addition to institutional differences, state features such as physical size and population may also exacerbate logistical difficulties involved in regulating polluters. For example, in states like Texas or California, that are about 268,596 and 163,696 square miles respectively, reaching each "navigable" body of water and each entity subject to state/federal environmental statutes is a logistical nightmare. Environmental agencies often lack the staff and resources necessary to compensate for the scope of day-to-day regulatory requirements. Thus, these factors may also be powerful motivators of enforcement behaviors, such as inspections and violation assignments (which are dependent upon inspections being performed in the first place).

Lastly, a *state's willingness* to enforce environmental protection affects enforcement actions. We often think of states as being purveyors of developmental policies (as opposed to redistributive policies). In competition with each other for industry, state and local governments may set less-stringent environmental standards than does the federal government in hopes of attracting business" (Koontz 2002, 9; Moe 1989). The dependence of state economies on an industrial sector or natural resource may make states more reluctant to take on the role of strict regulator. And, we can see this reluctance reflected in the attitudes of environmental bureaucrats. Konisky's (2008) evaluation of his 2005 State Environmental Managers survey reveals that state regulators "believe that environmental regulations matter to industry investment decisions, and they perceive that concerns about the effects of environmental regulation on industry do, at times, lead their agency to ease their regulatory effort" (2). Thus, depending upon state economic conditions and dependence on various industrial sectors and natural resources, a state's environmental agency may be more or less willing to take enforcement actions.

Of course, these pressures extend beyond the bureaucracy to state legislatures, executives, and courts, as well. Although environmental bureaucrats directly perform enforcement activities, state-level political principals wield influence through appointment powers, adjudication of disputes, and appropriations and reorganization powers. Current literature strongly supports that political and special interests drive regulatory preferences and behavior, with more liberal (Democratic) states pushing for more aggressive regulation, and more conservative (Republican) states pushing for leniency or more cooperation with industry (Hays et al. 1996; Daley and Garand 2005; Woods 2008; Ringquist 1993; Bacot and Dawes 1997; Davis and Davis 1999; Wood 1992; Atlas 2007). And, these political and special interests are considered an extension of citizen interests and environmentalism, which also shape environmental protection actions. States with more environmentally minded citizens and political officials tend to prefer (or at least tolerate) more aggressive regulatory actions.

Political control and state environmentalism are often cited as the primary motivator of environmental policy behavior in the states. This is notable because—as I point out above—this argument relies on the assumption that state regulators are acting as the hands of political principals, simply translating principals' preferences into day-to-day enforcement actions. However, there are reasons to question the strength of that assumption. To start, as Konisky (2008) points out, bureaucrats hold individual preferences and attitudes related to regulation that can and do affect their behavior. Additionally, bureaucracies, like all organizations, are made up of unique organizational cultures, including rules, norms, and traditions that drive employee decision making. This exists apart from and within the constraints of political pressures. Thus, in order to understand state-level regulatory actions and the consequences of those actions, we must consider how the organizational characteristics of environmental agencies strengthen, interrupt, or even eliminate political pressures. To do this, we must first identify how environmental agencies differ across the states.

STATE AGENCY DESIGN CHOICE

In 1970, when states were told that they needed to comply with new federal regulations that required them to create implementation plans and issue enforcement of those plans, states had to make a number of organizational decisions related to how they would address these new mandates. States had to decide if they would continue to house environmental protection efforts where they were currently (and ineffectively) implemented—in public health and natural resource agencies—or whether they would follow the lead of the EPA by creating a single-mandate agency dedicated to environmental protection. This was a difficult decision for two reasons. First, states understood that the creation of the EPA was related to the failure of state institutions and policies to protect clean air and water. In states where citizens were subject to disastrous environmental consequences, maintaining any kind of status quo would have likely been politically unpopular. However, the creation of a new agency would be costly and would challenge political officials who promised to eliminate wasteful government redundancy and prevent government overreach. Existing literature supports that these factors likely played a role. For example, in 2013, Sinclair and Whitford evaluated the factors that determined how states organized their environmental protection and public health mandates. Sinclair and Whitford (2013) identify two main factors that determined the choice states made to create a separate environmental agency: state liberalism and the prevalence of environmentally related illnesses, including congenital malformations and respiratory cancer. For more liberal states, fear over large and overbearing government and fear of the counterculture (including the growing number of environmental activists) was less pronounced. Thus, it is not surprising that more liberal states may have demanded—or at least not fought—the creation of mini-EPAs that would make environmental protection efforts most visible and could potentially protect the agency from the influence of other interested parties, such as natural resource regulators and agriculture. Additionally, as I note above, deteriorating health due to pollution highlighted the failures of state agencies and boards mandated to protect the environment.

Unfortunately, aside from Sinclair and Whitford's analysis, we do not have much information about the debates that states engaged in, in deciding how and why to organize their environmental protection agencies as they did. For example, in looking at the history of Texas's transition from an air and water board controlled by the Department of Health to an independent air board that acted as the state's environmental protection agency, neither legislative journals or newspapers from 1973 account for the organizational change with any fanfare. Like many state environmental agencies (e.g., Connecticut, Rhode Island, Washington, Arizona, etc.), Texas's environmental agency describes the creation of their environmental protection agency as being the consolidation of a number of independent water, air, and natural resource boards and commissions into a single environmental protection agency managing all programs. Although agency reorganizations are often contentious and laborious policy changes, state environmental agencies that provide insight into the history of their agencies, speak of these consolidations or relocations as if they happened out of necessity and without much conflict. It is difficult to challenge this narrative without news reports or legislative accounts to provide additional information. Thus, it is hard to know beyond ideological concerns and environmental health statistics, what may have driven states to choose and—importantly—keep certain organizational structures or what arguments were made for or against existing structures (Sinclair and Whitford 2013).

Regardless of motivation, states made one of two organizational choices: create a new environmental agency (often out of the many parts of environmental protection efforts spread throughout various commissions and boards) or invest new mandates in existing public health or natural resource agencies, commissions, or boards (where some environmental protection mandates were already being carried out). However, since then, many states have chosen to reorganize their environmental protection efforts since the 1970s (e.g., Michigan, Indiana, Pennsylvania, Utah, Oklahoma, and others). Currently, environmental protection is executed by fifty environmental agencies that are structured in three main ways: (1) as mini-EPAs; (2) as combined public health and environmental protection agencies (PHEPs); or (3) as combined natural resource conservation and environmental protection agencies (NREPs) (Hopper 2017; Hopper 2019). As depicted in figure 1.1, environmental agency organizational types differ across the states. In thirty states, environmental protection policies are implemented by state-level agencies that are organized to mirror their federal counterpart, the Environmental Protection Agency (EPA), where environmental protection is the sole purpose of the agency. However, in fifteen states, environmental protection policies are implemented by agencies that implement both environmental protection and natural resource conservation programs. And, in five states, environmental protection policies are administered as a part of public health agencies. Although public health and natural resource conservation are complementary policy areas to environmental protection, the choice to combine environmental enforcement activities with other activities, such as the management of

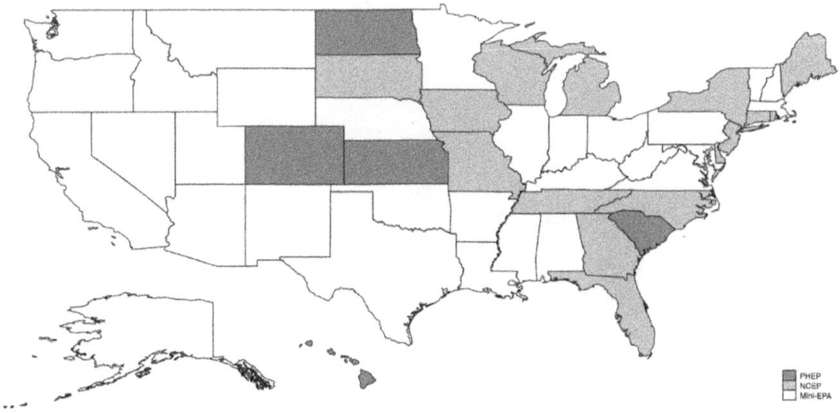

Figure 1.1. Environmental Agency Design Types by State.
Created by author, using Stata.

natural resource extraction and revenue, the management of state parks and wildlife, vaccine administration, testing for sexually transmitted diseases, and the regulation of marijuana markets has meaningful implications for day-to-day enforcement decisions (Hopper 2017; Hopper 2019).

Chapter Two

A Theory of
Environmental Agency Design

Organization is not a neutral factor. This was a primary concern of the Ash Council, in issuing their recommendations to Nixon regarding the organization of environmental protection programs. The memo sent to Nixon on April 29, 1970, notes that "while good people can sometimes make a poor organizational form work, more to the point is the fact that the system within which people operate can make it difficult for them to reach their institution's objectives." The council stated definitively that organizational choices were a "major determinant of the success of almost any enterprise." This is true. Organizational choices determine who is in charge, how objectives and tasks flow from one individual to the next, and how employees interpret their primary goals and the means appropriate for achieving those goals.

As I note in chapter 1, a lot of thought went into deciding how environmental protection programs should be organized in the United States—at least at the federal level. States already had their own environmental programs that varied in their organization, norms, and goals. Had the federal EPA become both creator and primary executor of environmental enforcement, these organizational differences may not be as consequential. However, since the partial preemption system relies on the states to enforce federal environmental standards, differences in organization among state environmental agencies continue to produce salient implications.

THE IMPLICATIONS OF AGENCY DESIGN CHOICE

Environmental control in the American states is complex. There is "no uniformity in the way states are organized to implement . . . major federal environmental statutes" (Burke et al. 1995, iv). The combination of public health and natural resource conservation with environmental enforcement in twenty states is a reflection of this lack of uniformity. However, the combination of tangentially related policy areas within a single bureaucratic agency is not uncommon. Combining policy areas ensures that "groups with similar missions and frequent working relations are grouped together . . . avoid[ing] conflicts of jurisdiction and permit[ting] far greater economy and efficiency in government" (Denhardt 2011, 60); the reorganization of agencies is often justified by possible improved coordination and lower costs. In addition to allowing political officials to address perceived bureaucratic inefficiency and waste, agency design choices have powerful implications for how much control political principals are able to wield over policy implementation. "Calculations about the 'proper' design of administrative agencies are shaped less by concerns for efficiency or effectiveness than by concerns about reelection, political control, and, ultimately, policy outcomes . . . some structural arrangements allow more control by political actors than others do" (Lewis 2004, 3). The organization of a bureaucratic agency, including its governance structure, term limits, and appointment process can matter a great deal in determining whose political priorities are expressed via policy implementation.

In addition to these factors, though, decisions made about which programs or departments to combine together under a single agency head also pose compelling ramifications. Take, for example, Cohen et al.'s (2006) evaluation of the creation of the Department of Homeland Security (DHS). In 2002, following the terrorist attacks on the World Trade Center and Pentagon on September 11, 2001, DHS was created by combining twenty-two agencies, including the Immigration and Naturalization Service, Federal Emergency Management Agency (FEMA), Environmental Measurements Laboratory, Energy Security and Assurance Program, and the U.S. Coast Guard, into a single department. The common rationale I reference above was used to justify this combination: it would increase coordination between agencies handling areas of security—coordination which was lacking in the lead-up to September 11. However, there were a number of warnings from those in the Bush administration regarding the creation of a massive, multi-mandate agency dedicated to homeland security initiatives (Cohen et al. 2006, 739). The authors argue that the transfer of a number of agencies under one umbrella homeland security agency resulted in two consequences: (1) agencies were forced to transfer resources from their original or legacy mandates to homeland security goals and priorities, and (2) "the new organizational control allowed the Administration to downplay the portions of the organization that remained focused on the legacy mandates" (739). This focus of resources and efforts onto "homeland security" goals can be seen in DHS's and FEMA's response to Hurricane Katrina in 2005. A number of experienced FEMA employees left the agency after its transition into DHS and the "merger accelerated a process through which FEMA's natural disaster and mitigation missions were eviscerated" (Cohen et al. 2006, 740). The organizational choices made in the creation of DHS led to the focus on some goals and mandates at the expense of others, and that refocusing of goals had tangible consequences.

That the consolidation of multiple mandates within a single agency led to difficulties for some mandates in the case of DHS is not surprising given the existing scholarship on agency combinations, goal conflict/ambiguity, and competing mandates. In 1989, Wilson criticized "multi-task conglomerates," stating the following:

> People cannot easily keep in mind many quite different things or strike reason-able balances among competing tasks. People want to know what is expected of them; they do not want to be told, in answer to this question, that "on the one hand this, but on the other hand that. . . ." No single organization . . . can perform well a variety of tasks; inevitably some will be neglected. . . . Running multi-task conglomerates is as risky in the public as in the private sector. (371)

Wilson's critique of "multi-conglomerate" agencies is that they invite confusion and conflict for bureaucratic workers; this happens because organizational goals "lose clear meaning and become ambiguous when [they] invite a number of different interpretations" (Chun and Rainey 2005, 531). This is particularly likely to be the case if the combined agencies are tasked with competing mandates.

For example, the defunct U.S. Minerals Management Service (MMS) combined regulatory functions, such as regulating offshore drilling, and nonregulatory functions, such as "facilitat[ing] the use of federal offshore energy resources to secure U.S. energy independence" (Carrigan 2017, 15). After the Deepwater Horizon explosion and spill in the Gulf of Mexico, many pointed to these competing mandates as problematic, given the potential for encouraging a cozy relationship with regulated entities and the capture of the agency by industrial interests. This capture was made possible because one of these mandates at MMS was more socially and politically palatable; focusing on revenue collection and the use of energy resources was a more rewarding set of goals and tasks than regulating offshore drilling (Carrigan 2017). When two mandates are combined, politically unpopular regulation and enforcement lose out.

However, even when two regulatory activities are positioned against one another, it is likely that one of those sets of regulatory activities receives more focus and resources than the other (Chun and Rainey 2005; Perry et al. 1999). First, as with workers in any organization, when given multiple tasks, bureaucrats must decide how to allocate their time among those tasks. Incentives to perform tasks differ, however, so time and efforts are unlikely to be distributed equally (Dewatripont et al. 2000). Additionally, Gilad's (2015) analysis of the effect of goal ambiguity on the attention to and prioritization of financial regulation tasks suggests that simply combining mandates—regardless of what they are—can lead to the development of different priorities within segments of a department or agency. These differences in priorities create discord within the agency, which opens space for outside interests to permeate the agency and gain undue influence over the direction of an agency's policy implementation. Additionally, two agencies that are charged with the same task—for example, two state environmental agencies tasked with protecting their individual state's environment—could end up with completely different organizational identities, cultures, and enforcement styles/preferences because of an agency combination (Gilad 2015). This occurs because agency combinations combine two sets of unique organizational preferences, norms, and rules. That is to say, in combined agencies, there are multiple, competing cultures that define the purpose and appropriate methods of implementation differently. Thus, the organizational culture that emerges from a combination is often the result of a push and shove between two competing organizational identities.

So, what might this mean for environmental protection agencies, where public health and conservation goals and tasks compete with day-to-day environmental enforcement activities? Previous work spells out some of the differences in organizational identity between public health, natural resource conservation, and environmental protection approaches to enforcement (Hopper 2017; Hopper 2019). Although these areas of policy are often thought to be complementary and potentially even redundant—thus, the justification for their combination—regulatory preferences and behaviors differ between public health, natural resource conservation, and environmental protection officials. And, those preferences differ in a way that affects how states protect the environment.

Public Health vs. Environmental Enforcement

"Environmental health was one of the very earliest of organized public health activities" (Kotchian 1997, 246).[1] For example, in the United States, local public health organizations formed in response to water-borne illnesses, air pollution, unsafe food production and consumption, and sewer infrastructure problems spurred by the industrial revolution and rapid population growth. These public health organizations held primacy over a variety of early environmental protection efforts until the 1970s, when growing demand for protection from environmental exposure led to the creation of environmental statutes focused heavily on industrial regulation.

For nations, such as the United States, this focus on industrial regulation was thought to require a new venue for policy implementation because American public health agencies had long been thought to consider environmental health a secondary task. As Gordon (1991) describes, "there is a serious deficit in interest and support for environmental health objectives among public health professionals" that dates back to even before the creation of single-task environmental protection agencies. These public health professionals were more interested in research and saw regulatory air and water programs as not public health programs, but, rather, "just regulation" (Gordon, 1991, 6).

And, as public health agencies have continued to evolve, the distance between public health agency priorities and environmental protection priorities has continued to grow. In the decades since the environmental movements of the 1970s, public health agencies have become much more active and involved in providing limited health care services, regulating hospitals, and partnering with Medicaid to reform payment and delivery services (see ASTHO, 2016, for example). With the passage of more extensive health care policies and programs (e.g., the U.S. Affordable Care Act in 2010), public health agencies have helped to establish insurance exchanges and to even provide healthcare services directly. This transition toward health-care liaison or provider can be seen in the budgets of American states. For example, in Kansas, health care finance now accounts for just short of 90 percent of the Kansas Department of Health and Environment's (KDHE) expenditures. Environmental expenditures account for less than 3 percent of the agency's budget (KDHE, 2013). Although the growth of health-care finance expenditures does not necessarily indicate a significant decrease in environmental spending (i.e., Kansas has always spent less on environmental protection than health), the overwhelming amount of agency funds going toward the health-care finance program signals its importance within the agency.

As I note above, public health agencies *do* perform regulatory functions, including "administering public health laws, sanitary codes, licensing programs, administrative rules and regulations, and the inspection and approval of environmental and health related facilities" (American Public Health Association 2016); however, we do not often think of public health agency workers as regulators, in the same way that we think of environmental enforcement workers. For public health agencies, regulation is one of a variety of tasks, which include non-regulatory educational programs, professional training, and disease control. Furthermore, the backgrounds of public health officials and environmental enforcement officers or inspectors differs, as environmental agencies generally employ "technical specialists," such as hydrologists or engineers, while public health agencies look for individuals with a background in epidemiology.

Thus, meaningful differences exist within the backgrounds and priorities of those working together within PHEPs. And, this diverse set of workers is expected to implement an equally diverse set of programs and policies. The question, then, is how might this complexity within the agency affect the way PHEPs execute environmental enforcement? As I note above, it is possible that the numerous goals and mandates within these agencies, alone, may lead to the neglect of environmental protection goals; however, the differences I have described thus far between public health and environmental protection may also affect the regulatory preferences of the agency.

As defined by the existing literature, regulatory style describes the stringency with which bureaucratic personnel choose to enforce the law (Shover et al., 1986; Hunter and Waterman, 1992; May and Winter, 2000). Enforcement may be executed without much discretion or stakeholder involvement or it may be more flexible and aimed at collaborating with stakeholders. Public health agencies have traditionally abided by the recommendation that stakeholders (e.g., elected officials, community members, scientists, and regulated parties) should be at the center of any process for understanding environmental health problems (Omenn, 1996). Stakeholder involvement helps to ensure that we accurately approximate risk and are able to develop the most cost-effective options (Omenn 1996).

This focus on cost-effectiveness reveals an important difference in the statutory language in policies aimed at public health and environmental protection agencies. For public health agencies, the rising cost of health-care services has led to a renewed focus on ensuring that public health solutions consider costs, weighing policy alternatives to determine which option will provide us with the greatest public health protections at the lowest cost (see, for example, Goldie 2003). Conversely, some of the most powerful environmental statutes, such as the U.S. Clean Air Act, have been interpreted by Congress as policies for which costs should not be considered when setting standards (e.g., 1981–1982 Senate Subcommittee on Environmental Protection rulings on air quality standards). One piece of the logic behind excluding costs from the environmental protection equation is that the consideration of costs may lead us to incorporate polluting industry's preferences into the regulatory process, and that this could potentially lead to ineffective regulation. Thus, the stakeholder involvement that is fundamental to environmental program success from a public health perspective is discouraged or even directly removed from the regulatory process when executed by single-task environmental protection agencies, such as the EPA (Omenn 1996).

Lastly, as I describe in my analysis of interviews with public health officials in chapter 4, the funding mechanism for public health activities serves to create an adversarial relationship with the EPA. While environmental protection activities are primarily funded by federal grant dollars from the EPA, most public health services and programs are funded through state or local taxes or fees (e.g., charges for licenses or permits). For environmental protection programs, the funding stream from the EPA creates a financial dependence that helps the EPA to drive the behavior of state-level environmental officials. This dependence is less consequential for public health officials, who are more beholden to state and local political pressures. Thus, public health officials may be resistant to EPA pressure to prioritize certain programs or to execute regulation in a particular way. Simply put, this means that the differences introduced into environmental protection—a growing focus on providing health care services and the prioritization of stakeholder inclusion and cost-benefit analysis—are likely not abated by the EPA's influence. That said, discrepancies in regulatory preferences between the EPA and natural resource agencies may be even more pronounced.

Natural Resource Conservation vs. Environmental Enforcement

In the mid-1800's, it became clear that we had exhausted a number of natural resources, including the wild bison that dominated the American plains, birds, timber, and healthy soil.[2] In response to what many saw as a decimation of the idealistic American landscape a push toward preserving or conserving those resources grew. The National Park System was developed, better farming practices were encouraged and implemented, and restrictions on the hunting and trade of animals were enforced. States (and the federal government) created boards or agencies to create and enforce these new conservation standards. It was in the economic interest of states to create programs that would prevent a "tragedy of the commons." Many of these bureaucracies were the predecessors to more modern state environmental agencies that focus on pollution control. Natural resource conservation programming and environmental protection programming mirror each other in their attempts to control certain types of behavior, such as the exploitation of natural resources or the excessive release of toxic emissions, respectively. However, while the two policy areas are rooted in regulation, the focus and purpose of regulatory programs are fundamentally different.

Natural resource conservation typically includes the maintenance of state parks, lands, and animal populations, along with the control of state resources, such as oil, natural gas, minerals, and timber. Conversely, environmental regulation focuses on the control of air and water pollution. There is some overlap between mandates, in the conservation of wildlife; however, states tend to focus much more on controlling wildlife for the purposes of hunting, while environmental regulators—such as those at the EPA—are more focused on the maintenance of healthy eco-systems (Koontz 2002). This contrast in the approach to protecting wildlife helps illustrate an important difference between natural resource conservation and environmental protection agencies, in how they define the *purpose* of regulation. For natural resource conservation agencies, regulation is intended to protect valuable contributors to the economy—as a continued wildlife population may be an economic contributor, in regard to tourism, sport, etc. This also would apply for timber, natural gas, minable ores, and land. The management of these resources often results in tangible, material goods for state governments and state residents. In the Pacific Northwest, for example, "every million board-feet of timber sold is linked to a dozen local jobs annually," and local community infrastructure and education may be funded by portions of revenue coming from timber taken from public forests (Koontz 2002, 36; Satchell 1996; Gilless et al. 1990). Thus, state-level natural resource conservation involves some consideration of both how to protect resources and how to protect the industries that make those resources economically fruitful. In fact, several states (e.g., Indiana, Washington, and Oregon) have conservation laws in place that actually prioritize economic outcomes over environmental outcomes; state legislation specifies that the primary purpose of forest management is to ensure that financial returns are earmarked to fund education (Koontz 2002; Souder and Fairfax 1996). Conversely, environmental regulators are often asked to regulate the behavior of industries, with no guarantee that regulation will not disrupt an industry's economic contributions to the state. In fact, many of regulators' options for punishment involve ensuring that an industry cannot make money via exploitation or "breaking the rules."

Consequently, natural resource conservation agencies' preference for regulation must include more flexibility or willingness to compromise and collaborate with industry stakeholders. This type of regulatory style has been labeled the negotiated compliance model, in which there is "a dominant orientation toward obtaining compliance with the spirit of the law through the use of general, flexible guidelines" (Shover et al. 1986, 11; Hunter and Waterman 1992). Under this model of regulation, agency personnel engage with a regulated industry and take "an accommodative stance toward the regulated" (ibid.). However, the adoption of this model of regulation is not just a preference; it is actually required by many laws that involve conservation mandates. For example, the federal laws that direct natural resource conservation agencies, such as the Federal Land Policy and Management Act, the Endangered Species Act (Habitat Conservation Plan), and the National Forest Management Act require that agencies engage in a "planning process" with regulated entities—or, what we might call management-based regulation. Management-based regulation approaches the regulatory process by asking "firms . . . to produce plans that comply with general criteria to promote [a] targeted social goal" (Coglianese and Lazer 2003, 2). These plans could include criteria that instructs employees how to identify problematic procedures or policies, how to monitor problems, how to create and implement training procedures, and how to evaluate progress. Through the creation of these plans, industries become an active stakeholder in their own regulation.

In addition to helping to protect valuable economic contributions from resources and extractors, these laws also help to ensure that conservation efforts are carried out efficiently and effectively. Successful conservation efforts rely upon the realization of "the synergistic effects and complex interdependencies among the various component and stressors that make up [an] ecosystem" (Karkkainen 2001, 947). This means efforts must include a number of stakeholders, industries, and agencies, in order to piece together the very complex set of variables that may affect the health of ecosystems. Another implication of this complexity is that the continued health of ecosystems is difficult to predict; therefore, management of an ecosystem must be flexible enough to evolve when new information is uncovered.

Thus, industry-inclusive and flexible planning processes play a fundamental role in regulation for conservators of natural resources—an inclusivity and flexibility that has not always been embraced by environmental protection's command and control approach, where "agency personnel should employ strict sanctions, such as fines and penalties" (Hunter and Waterman 1992, 404). For example, at the federal level, the Department of Interior has "tended to favor development and resource extraction interests" as opposed to environmental groups, and the agency is less "regulatory-minded" than the EPA or the Occupational Safety and Health Administration (OSHA) (Stewart 2001, 35). DOI and the U.S. Forest Service have both openly expressed frustration with what they argue are prohibitively strict enforcement measures implemented by the EPA (USDA 2002; Nie 2008). In addition to expressing frustration, these conservation-minded agencies have also revised some environmental regulations to include more flexibility, particularly as it relates to oil and gas exploration, federal grazing, or development (for example, see Luther 2006; Humphries 2004; 66 Fed. Reg. 54, 834 2001). This resistance occurs as the EPA continues to argue in favor of strict enforcement against noncompliant industries (GAO 2000; 2006).

In sum, both public health and natural resource conservation approaches to regulation, which often focus on reducing or controlling cost and including stakeholders, differ from an environmental protection approach that may downplay or outlaw considering cost (e.g., the Clean Air Act) and takes a command and control approach of setting and enforcing standards without direct input from industry or individualized plans for polluters. Thus, when a public health or natural resource conservation agency is tasked with enforcing environmental regulations, they must find a way to balance between these competing approaches. However, competing approaches and mandates lead to the prioritization of some approaches and mandates over others. In the section that follows, I lay out the reasons why we might expect for a negotiated compliance approach to be prioritized by public health and natural resource conservation agencies mandated with protecting the environment.

PUBLIC PRESSURE AND REGULATORY ACTIONS

In an analysis of the downfall of the Minerals Management Service (MMS), Carrigan (2014) writes that the agency took on competing mandates: collecting and distributing revenue generated through "onshore and offshore leases of federal property to companies who used the lands to extract oil and gas for private sale" and "overseeing the orderly development and regulation of offshore oil and gas production on the Outer Continental Shelf." It was these competing mandates that critics of the agency pointed to as a motivation for regulatory capture, in which the agency's need to protect industry and maintain a cooperative relationship led to laxer regulation and enforcement. These are similar to the competing mandates that exist for many combined agencies that are expected to both collect fees and revenue from industry, while also regulating the activities that may boost those fees and that revenue. Furthermore, it is likely that the pressure introduced by competing mandates such as these are even more pronounced at the state level, where economic pressures are closer and more personal to regulators. Konisky (2008) notes that 71 percent of surveyed environmental workers believe that their regulatory decisions may affect whether or not industries locate in their state; another 55 percent think that relaxing regulatory standards would have a positive effect on their state's ability to compete with other states in attracting industry (330–331). This finding supports the argument that even non-elected officials are subject to public pressure that may incentivize them to engage in more cooperation and flexibility with regulated entities (Koontz 2002).

This adds an interesting layer to the conventional wisdom that state legislators and governors are the primary applicants of pressure for regulators. The lack of electoral consequence could increase protection from direct citizen pressure; the power to persuade would ultimately be in the hands of those political principals acting on citizen interests. But—at least in the states—this does not appear to be the case. Regulators are living and working within the economies that they are expected to regulate. As a regulator, if I am cognizant and concerned about the effect my regulatory actions may have on my friends and neighbors, it is not surprising that I may choose to be more flexible with industry when I have the option. And, in PHEPs and NREPs, I do have that option.

In addition to this pressure, regulatory solutions that take a negotiated compliance approach, such as management-based regulation or voluntary compliance, are quickly becoming acceptable methods of regulation and enforcement, helping to supplement the more traditional command and control efforts taken by the EPA (see Coglianese and Nash 2001; Bennear 2006, for example). Thus, the shift toward more collaborative and flexible approaches with industry would not be completely out of line with the regulatory "toolbox" used by agencies like the EPA (Bennear 2006). Therefore, the door is already cracked for using more industry-inclusive approaches; making environmental enforcement part of an agency that already prefers negotiated compliance swings that door wide open.

THE MEDIATING IMPACT
OF STATE INSTITUTIONAL CAPACITY

In addition to indirect public pressures, are the more direct pressures imposed on bureaucrats by political principals, such as state legislatures, governors, and courts. As I note previously, we often associate environmental policy behavior with these institutions, where policy decisions are initially made and, then, delegated to bureaucrats for implementation. Additionally, we may think of bureaucratic agencies as the "hands" of political principals, simply carrying out principals' policy preferences. However, the relationship between principals and bureaucratic agencies is more complex. As I discuss in chapter 1, bureaucratic behavior is driven by a number of external *and* internal factors. The extent, nature, and success of outside pressure is subject to an agency's preferences, how easily elected officials can withhold funds and direct programs, and elected officials' and citizens' perception of the bureaucracies' capacity and trustworthiness (Carpenter 2001; Potoski and Woods 2002; Wood and Bohte 2004). More simply, this means that which institution(s) wields the most control over the direction of environmental policy is determined by the capacity of the various institutions fighting for power. In the states, legislatures, governors, and bureaucracies differ significantly in their institutional capacities (e.g., professionalization, term limits, veto powers, pay, turnover, etc.). Thus, factors like agency design or political control likely matter more or less depending upon state institutional capacity.

A THEORY OF ENVIRONMENTAL AGENCY DESIGN

Looking at table 2.1, we can see that the enforcement behavior of environmental agencies differs significantly across the states, including stark differences in inspection rates, monetary penalties, and enforcement assignments. Table 2.2 shows how the states rank against one another, using these variables. As I have discussed in this chapter, this variability is often explained by looking to the economic and political landscape of a state: how is the economy doing? Who is in control? But, the existing literature has neglected to consider those making the decisions on the ground: environmental regulators, themselves, and the agencies that shape their day-to-day preferences and behavior, along with how bureaucracies' power may be shaped by the capacity of state institutions. In the previous pages, I have made the following four arguments about environmental agency design: (1) any combination of policy areas within a single agency allows for some tasks and mandates to be prioritized over others or for parts of the agency's programming to be neglected; (2) public health and natural resource agencies have their own regulatory preferences that often include prioritizing the inclusion of industry stakeholders, considering cost, and focusing on education and planning; (3) the economic, social, and political pressure placed on state-level bureaucrats uniquely positions them to favor the regulatory preferences embraced by public health and natural resource agencies, particularly as these approaches are growing more common within environmental regulation and when funding and implementation structures exist that bring regulators closer to citizens (e.g., with the traditional organization of public health offices and programs). Finally, (4) the strength of agency design in shaping day-to-day enforcement decisions is dependent upon state agencies' capacity and the capacity of those institutions with which agencies compete for influence.

In the chapters that follow, I test these propositions by looking closely at each agency type and speaking to those that work within them. I also use agency documents and enforcement data to establish the preferences of these agencies and how those preferences translate into regulatory actions. What I find is that the exclusion of the bureaucracy from the story of environmental

Table 2.1. Environmental Enforcement Data by State (Average 2010–2017)

	Annual Penalty Assignment per Number of Polluting Facilities	Number of Enforcement Actions Taken per Number of Violations Discovered	Number of Inspections Performed per Number of Polluting Facilities
Alabama	504.15	1.37	0.87
Alaska	793.15	3.62	0.34
Arizona	1673.92	2.82	0.40
Arkansas	326.06	2.09	0.70
California	1705.90	5.17	0.65
Colorado	992.84	2.17	0.24
Connecticut	512.16	2.34	0.26
Delaware	2394.85	0.84	0.44
Florida	297.90	0.59	0.43
Georgia	215.01	1.09	0.29
Hawaii	531.58	1.07	0.55
Idaho	195.58	3.82	0.13
Illinois	941.77	1.41	0.30
Indiana	748.66	1.95	0.43
Iowa	11.81	1.06	0.22
Kansas	333.89	2.43	0.59
Kentucky	1032.86	7.07	0.53
Louisiana	1609.70	8.53	0.22
Maine	372.88	0.59	0.33
Maryland	666.54	2.29	0.39
Massachusetts	872.21	0.91	0.16
Michigan	474.05	0.86	0.33
Minnesota	905.06	6.44	0.28
Mississippi	613.86	1.29	0.35
Missouri	70.61	4.00	0.54
Montana	587.17	1.55	0.20
Nebraska	264.80	2.96	0.38
Nevada	161.24	3.21	0.27
New Hampshire	329.49	4.73	0.26
New Jersey	929.24	11.82	0.29
New Mexico	1518.79	2.65	0.12
New York	175.06	1.09	0.10
North Carolina	181.02	1.01	0.94
North Dakota	32.77	0.06	0.12
Ohio	552.63	3.77	0.31
Oklahoma	222.08	0.50	0.16
Oregon	542.05	1.77	0.46
Pennsylvania	2017.52	4.89	0.60
Rhode Island	504.04	2.21	0.27
South Carolina	451.09	1.02	0.53
South Dakota	17.58	0.32	0.73
Tennessee	319.38	1.99	0.80
Texas	2569.48	1.04	0.25
Utah	135.78	0.92	0.54
Vermont	1.97	3.41	0.42
Virginia	2322.71	1.59	0.27
Washington	2438.26	9.13	0.68
West Virginia	2146.37	0.93	0.51
Wisconsin	1013.64	0.61	0.31
Wyoming	2116.23	3.12	0.33

Table 2.2. Environmental Enforcement Rankings by State (Average 2010–2017)

	Annual Penalty Assignment per Number of Polluting Facilities	Number of Enforcement Actions Taken per Number of Violations Discovered	Number of Inspections Performed per Number of Polluting Facilities
Alabama	28	31	2
Alaska	19	12	26
Arizona	9	17	21
Arkansas	35	24	5
California	8	6	7
Colorado	14	23	41
Connecticut	27	20	39
Delaware	3	44	17
Florida	37	47	19
Georgia	40	34	32
Hawaii	26	35	10
Idaho	41	10	47
Illinois	15	30	31
Indiana	20	26	18
Iowa	49	36	43
Kansas	33	19	9
Kentucky	12	4	13
Louisiana	10	3	42
Maine	32	46	28
Maryland	21	21	22
Massachusetts	18	42	46
Michigan	30	43	27
Minnesota	17	5	34
Mississippi	22	32	24
Missouri	46	9	12
Montana	23	29	44
Nebraska	38	16	23
Nevada	44	14	35
New Hampshire	34	8	38
New Jersey	16	1	33
New Mexico	11	18	49
New York	43	33	50
North Carolina	42	39	1
North Dakota	47	50	48
Ohio	24	11	29
Oklahoma	39	48	45
Oregon	25	27	16
Pennsylvania	7	7	8
Rhode Island	29	22	36
South Carolina	31	38	14
South Dakota	48	49	4
Tennessee	36	25	3
Texas	1	37	40
Utah	45	41	11
Vermont	50	13	20
Virginia	4	28	37
Washington	2	2	6
West Virginia	5	40	15
Wisconsin	13	45	30
Wyoming	6	15	25

regulation in the American states is consequential; the choices made about how to organize environmental agencies matter. And, these choices continue to matter in the face of political and economic pressures. Organizational decisions have long-term consequences. Nixon's advisors knew it when shaping the EPA, and, in the states, I argue, it is neither a neutral nor weak factor in determining the substance of environmental enforcement.

NOTES

1. See Hopper (2019) for a similar description of public health regulatory preferences.

2. See Hopper (2017) for a similar description of natural resource conservation regulatory preferences.

Part I

THE VALUES AND PREFERENCES OF ENVIRONMENTAL AGENCIES

Chapter Three

Environmental Enforcement in Combined Natural Resource Conservation and Environmental Protection Agencies

For many states, the conservation of state land and resources is a point of pride. Beautiful state parks that include opportunities for camping, hiking, hunting and fishing, and other recreational activities dot the American landscape, and many states rely upon the harvest of timber, the mining of ores, or the extraction of natural gas/oil to bring money into the state through exporting and the creation of jobs. The benefits of conservation (or at least the responsible management of state resources) are often noticeable and felt by communities who rely on natural resource jobs or the recreational opportunities provided by state parks and lands. Thus, it is not surprising that almost all states have an agency dedicated to conservation or natural resource conservation; however, in most states, this agency is not also in charge of environmental enforcement actions. Generally, these agencies deal only with state park or protected state land management, issuing hunting and fishing licenses permits, and controlling permits and fees for resource extraction.[1]

That being said, fifteen states in the United States currently combine natural resource conservation and environmental enforcement into a single agency with multiple mandates (Hopper 2017). As shown in the previous chapter, these states have some geographic clustering, with a handful in the Midwest and in the southeastern and northeastern parts of the United States, but they differ in regards to physical environment, dependence on natural resources, political control and ideology, and the presence and type of industry. For some of these states, environmental enforcement was added in the 1970s into an already existing agency dedicated to managing and conserving natural resources (e.g., New York), while for others (e.g., Tennessee), this decision was made later. As I have argued thus far, we should expect that this agency design feature currently adopted by fifteen states may affect the way these state agencies work to protect the environment. Specifically, due to the multiple (and potentially competing) mandates in the agency, along with the preference of conservators and political/community pressure to be more flexible and cooperative with industry, NREPs are likely to show a preference for more industry-inclusive behavior during the regulatory process.

To assess this proposition, in this chapter, I look closely at two NREPs—the Missouri Department of Natural Resources and the New York Department of Environmental Conservation. I have chosen these states for the following reasons. First, the difference in geographical location, environmental issues, and political landscape allow for us to consider how the agency design type may affect an agency within differing contexts. Additionally, both Missouri and New York adopted this agency type in the 1970s, and the structure of the agency has remained fairly stable, which allows for us to see some of the long-term effects of a decision that was made around forty years ago. In exploring the agencies' history, development, current stated goals and priorities, and public perception of the agencies, I evaluate how these agencies implement environmental enforcement. Following these cases, I look closer—through interviews with individuals working in NREPs—to determine how these organizational-level factors influence individual-level day-to-day behavior of environmental regulators.[2] In the discussion that follows, I illustrate how both cases and hours of interviews provide support for the logic that NREPs have adopted a particular regulatory style, and that style motivates enforcement behavior.

CASE 1: MISSOURI DEPARTMENT OF NATURAL RESOURCES

Missouri is a state with an abundance of natural resources, including fertile farmland; ores, such as iron, lead, coal, and limestone; and even a small amount of oil and gas near the state's borders. It is because of this abundance of natural resources that the state legislature created the Geological Survey of Missouri in 1853 to study Missouri's natural resources and to create a "thorough geological and mineralogical survey of the state so [the state] would know what was useful or valuable" (Balkenbush 2014, 12). The creation of the Geological Survey was Missouri's first step in establishing the administration of environmental protection programs and policies. Upon the evaluation of natural resources, Missouri also discovered that its public lands had substantial recreational value. This prompted the creation of the state park fund in 1917, another pivotal part of early environmental protection efforts in Missouri. Following the creation of the park fund, Missouri acquired land for its first state historic site—Arrow Rock Tavern in 1923—and its first state park—Big Spring State Park in 1924.

It would not be until 1943 that Missouri took its first steps toward establishing modern-day environmental standards, with the creation of the Soil and Water Districts Commission, which was intended to "further soil conservation practices on the farms of the state. The commission maintained oversight of the soil and water districts" (Balkenbush 2014, 10). This initial step, even though it focused on the health of water as well as soil, was still primarily focused on conservation efforts rather than pollution-control. Missouri's first *true* pollution-control efforts came in 1955 and 1958, as the state established its first solid waste law, and the Water Pollution Board was established within the Department of Public Health and Welfare. It is important to note two things about these preliminary pollution-control measures. First, the County Option Dumping Ground Law, Missouri's first solid waste law, was only adopted by 22 of the 114 counties in the states. Second, water pollution was dealt with exclusively within the arena of public health prior to the establishment of the MODNR in 1974. Thus, early pollution-control efforts were not particularly popular undertakings within the state, and they were treated as a separate mandate from the larger and better-resourced natural resource conservation efforts. This is a trend that would continue, even once environmental regulation and conservation efforts were officially combined.

In 1974, Missouri created the MODNR in response to federal environmental policy changes, a lack of quality drinking water, and growing problems with smog and hazardous waste (Balkenbush 2014). During this large agency reorganization, new environmental regulations, which focused on the control of pollution through the regulation of industry, were incorporated into historically strong natural resource and conservation programs that had been operating for over 120 years. Although an environmental regulatory apparatus existed in the state pollution boards, the state chose to relocate pollution control responsibilities into natural resource conservation programing, rather than to create a new agency dedicated to pollution control. This decision continues to shape the direction of Missouri's programming into present times.

A Lasting Legacy of Conservation and Flexibility with Industry

With the creation of the Geological Survey and the state park fund as Missouri's initial steps toward protecting the environment, it is clear that environmental protection in Missouri is rooted in the protection and preservation of lands and resources that Missourian's deem economically and aesthetically important.[3] This sentiment is reflected in current Missouri programming, as well. In MODNR's 2015–2020 Strategic Framework, a publication of

MODNR's goals, priorities, and plans for the next five years, the department states that it intends to achieve "its goals through partnerships, working with diverse interests from all parts of Missouri to support [its] mission and to better serve the public and manage [Missouri's] natural and cultural resources" (3). Partnership and cooperation are strong themes in Missouri's approach to environmental protection. MODNR's strategic plans consistently refer to the importance of "education," "outreach," and "communication," in "assist[ing] the department in building awareness and encouraging appropriate actions that result in a better quality of life for everyone" (8). Indeed, MODNR states the protection of health, well-being, and safety of Missouri citizens is a top priority, but the department actively relates these actions to "stewardship" of resources and the fostering of "collaboration and innovation within the department and with [the department's] stakeholders" (6). The focus of Missouri's environmental protection goals is on management through cooperation with industry, rather than through enforcement.

As I related in chapter 2, a strong emphasis on partnership and cooperation is a central component to the natural resource conservation approach to environmental protection. Therefore, it is not unexpected that MODNR, with historical roots in the preservation and conservation of natural resources and public recreational lands, would prioritize environmental protection approaches that emphasize partnership and cooperation with stakeholders, including polluters. This is important to note, as environmental regulation in Missouri is performed almost exclusively by MODNR, and, yet, the publicly embraced focus of the agency, as related by the agency, is the stewardship of natural resources (Balkenbush 2014). In support of this vision, Andrea Balkenbush, MODNR's chief of planning, wrote that "advances in science and technology will continue to help [Missourians] all be better stewards of natural resources. Because Missourians value their natural resources and the quality of life they provide, the Department of Natural Resources continues to fulfill its mission with broad public support" (13).

The embraced mission of natural resource conservation at MODNR that emphasizes flexibility with industry is also made clear by the fact that words such as "regulation," "control," "polluter," or "pollution" do not appear throughout the state's 2015–2020 strategic plan, except in reference to a goal to provide "compliance assistance to the *regulated* community," as "the department's enhanced compliance assistance program simplifies the *regulation* process and provides greater access to the resources necessary to help . . . customers thrive and succeed" (5). Any statements regarding enforcement are prefaced by MODNR's reliance on "compliance assistance tools," as the preferred option and the first step in response to pollution violations. To provide a more thorough example of these compliance assistance tools, we can look to the collaborative adaptive management approach MODNR used to address water quality issues in Missouri's Hinkson Creek. This approach helps to illustrate Missouri's dedication to using a natural resource conservation framework to address environmental problems.

Since 1998, Hinkson Creek, a twenty-six-mile stream that runs through the Columbia, Missouri area, has been on MODNR's list of impaired waters—meaning it does not meet EPA Clean Water Act standards. In 1999, the EPA assessed that the creek's impairment was due to stormwater runoff, bringing "insecticides, herbicides, chloride, heavy metals, and waste oil into the creek" (Ogden 2013). Shortly after this assessment, the Missouri Sierra Club and the American Canoe Association sued the EPA for failing to require MODNR to address the urban stormwater runoff problem. The city of Columbia, Boone County, and the University of Missouri countered the suit by claiming that the required 39.5 percent reduction in urban stormwater runoff suggested by the EPA would be "all but impossible to achieve" (ibid.). After a multi-year negotiation process, the EPA and MODNR agreed on a collaborative adaptive management approach to help clearly identify the root causes of the pollution and to assess possible solutions.

The collaborative adaptive management approach is a "science-driven, stakeholder-based process for decision-making . . . using a continuing process to make changes and then to determine the effect of those changes." It is a "science-based approach guided by a local stakeholder committee" (MODNR 2012, 1). An important component of this approach is the participation of local stakeholders, including industry representatives and environmental advocates. Working together with scientists, the stakeholder committee "represent(s) the community, recommend(ing) actions, monitoring, and modeling . . . for implementation" (4). The collaborative adaptive management process is a response to the EPA's earlier assessment of Hinkson Creek. The EPA suggested a large-scale cleanup that would have cost millions of dollars. MODNR, along with the city of Columbia and other local actors, challenged this assessment, requesting that they be able to perform their own set of analyses, using the collaborative adaptive management approach. This approach allows MODNR to assess the problem, while also considering the preferences of stakeholders. The collaborative adaptive management process extends flexibility and cooperation with local industry, making this approach a good fit for a natural resource conservation culture that embraces partnership with industry. The EPA's regulatory process does not explicitly prioritize the preferences of local stakeholders—something I discuss in detail in chapter 5.

MODNR proudly acknowledges the state's long history of protecting valuable state resources and lands. These environmental protection efforts, according to the state, help to support the economy and are the focus of much of MODNR's programming. Missouri's dedication to natural resource conservation approaches are apparent in MODNR's prioritization of stakeholder-inclusive approaches to environmental protection. MODNR's strategic plans and publications express the importance of education, cooperation, and flexibility, and the state approaches its environmental problems, using processes, such as the collaborative adaptive management process, to address environmental problems in the state. As the chief of planning for MODNR states, "even 161 years [after the creation of the MO Geological Survey], mineral resources and mining in [the] state are important natural resource protection issues and crucial to the [state] economy" (Balkenbush 2014, 12). And, this continued dedication to emphasizing the protection of natural resources has proved greatly influential in determining how environmental protection is approached. Enforcement efforts are not easily placed within the natural resource conservation framework put forth by MODNR.

Public Allegations of Leniency by MODNR

For MODNR, this focus on natural resource management, flexibility/cooperation with industry, and a definition of enforcement as an action of last resort, has brought sharp criticism over the years. For example, former MODNR employees have stated that the agency "lacks the 'political appetite' to strongly enforce state laws and [nurtures] a culture where employees are threatened with discipline if they talk to the media or the public" (Barker 2015). Additionally, an employee stated that MODNR has a "cozy" relationship with regulated entities, a relationship that does not provide a service to Missouri citizens. The *St Louis Post-Dispatch* reported that MODNR provided summaries of conference calls between MODNR, the Sierra Club, and Washington University's Interdisciplinary Environmental Clinic, along with other information, to Ameren, without Ameren filing a records request (Barker 2015). Ameren is a large holding company based in St. Louis, Missouri, for a group of electric, energy, and power corporations. Here is an example. MODNR was accused of providing more information to regulated entities than it is willing to provide to the public, giving the perception of collusion between regulators and polluters.

The *St Louis Post-Dispatch*'s investigative report is not the first time MODNR has come under fire for withholding information or colluding with regulated entities. In 2010, MODNR director Mark Templeton, appointed by Democratic governor Jay Nixon, resigned after MODNR delayed releasing water quality results that showed elevated E. coli levels in the Lake of the Ozarks, a large revenue source for the state's economy. Additionally, in 2007, *The Columbia Tribune* reported that more than four thousand pages of documents contained evidence that MODNR "levied civil penalties against large water-polluting animal farms, only to later reduce the penalties to about a quarter of the original amount" (Off 2007). Former employees and Missouri citizens referred to penalties as "lax" and stated that Missouri "is easy," in regard to getting away with violating environmental standards (Off 2007). MODNR's response to many allegations of leniency with industry are that "cuts in penalty amounts are a balancing act . . . while enforcing the state's environmental laws, [MODNR] also wants to help develop Missouri's economic development potential" (ibid.). Additionally, representatives from MODNR state that the agency always takes part in a negotiation process, working with industries to assess the parts of a problem that were truly the fault of the industry and to make sure that first-time offenders are given flexibility. These responses speak to MODNR's emphasis on flexibility and cooperation, supporting the agency's own words expressed in its annual reports.

CASE 2: THE NEW YORK DEPARTMENT OF ENVIRONMENTAL CONSERVATION

As with Missouri, New York has a rich history of protecting and managing natural resources, in particular the state's fresh water sources, forests, and estuaries. The state also manages 180 state parks, which cover around 350,000 acres (including historic sites). Efforts to protect New York natural resources began as early as 1880, when the state appointed "game protectors" (now, "conservation officers") (NYDEC, "History of DEC"). According to the New York Department of Environmental Conservation (NYDEC) (In quotes from news articles, some may refer to the agency using NYSDEC, whereas most use NYDEC), the appointment of these officers to enforce game laws actually preceded "the formation of the Division of State Police by 27 years." Five years later, the New York State legislature created the Forest Preserve of New York, which preserved lands in the Adirondacks and Catskills, and the forest ranger service. This was followed ten years later by the formation of the Fisheries, Game and Forest Commission, which dealt primarily with regulating hunting, fishing, and poaching prevention. In 1927, the Conservation Department was formed out of a number of commissions, and NYDEC points to this department as being one of the "forerunners of the NYDEC when it was formed in 1970" (NYDEC, "History of DEC").

Although it was forty-three years from the creation of the Conservation Department before New York combined some environmental health programs from its Department of Health with the Conservation Department and various commissions to create the NYDEC, the NYDEC claims to be one of the "first government agencies specifically formed for the purpose of overseeing all environmental concerns through one organization" (NYDEC, "History of DEC"). The state claims its department precedes the creation of the federal EPA and federal legislation such as the CAA. One of the new departments' first actions was to create an endangered species list in 1970, followed by actions to ban DDT, establishing park agencies, and taking action to protect freshwater wetlands (NYDEC, "History of DEC and Highlights"). These conservation actions were also accompanied by various environmental health actions, including addressing mercury pollution, lead-based paint poisoning, and the discharge of polychlorinated biphenyls (PCBs) (known to cause cancer). As state historical documents note, the department took on a significant number of tasks across the areas of health, agriculture, and natural resource management.

However, as is the case with Missouri, New York's environmental legacy is rooted in conservation. Most pollution control laws in New York were being handled at the city-level or by the state's health department prior to the creation of NYDEC; however, these laws, agencies, or commissions are not noted as the primary predecessors to the creation of NYDEC (NYDEC, "History of DEC"). NYDEC's description of its history and formation points specifically to conservation efforts, without reference to the pollution control efforts implemented by public health officials (NYDEC, "History of DEC"). In fact, in the NYDEC's document entitled "History of EC and Highlights of Environmental Milestones," the department points to 324 different milestones in the agency's history; 111 of those milestones are related to pollution control, while around 168 of those milestones are related to conservation efforts. And, although the state created an environmental agency prior to the creation of the EPA, the New York City smog episodes of the 1960s are considered part of the series of environmental disasters that led to the federal government moving the setting of environmental standards away from the states. State pollution control efforts (particularly in New York City) were failing—just as they were in other states. However, rather than create a new agency dedicated to pollution control, New York opted for the creation of an agency that would include all aspects of environmental protection, including its historically important Conservation Department.

Regulatory Cooperation, Education, and Collaboration in NYDEC

A focus on conservation and the regulatory approaches embraced within the conservation community continues to motivate NYDEC in the present day. This is reflected in the agency commissioner's yearly reports, in which the commissioner writes that the agency is "committed to protecting and preserving our shared environment and increasing opportunities for visitors . . . to enjoy our world-class natural resources" (NYDEC 2017, 1). Natural resource management accomplishments, including forest timber sales, partnerships for managing invasive species, arrests for hunting and fishing violations, expansion of recreational opportunities, development of artificial reefs, and conservation education take up thirteen pages of the twenty-five-page report in 2018 and sixteen pages of the twenty-eight-page report in 2017. Even in the remaining pages dedicated to pollution control, climate change, and "green" initiatives, the regulatory preferences of conservators to engage with regulated entities and work toward education and voluntary compliance are expressed explicitly. For example, in discussing the management of solid waste, the 2018 report details the funding of the New York State Pollution Prevention Institute, which "implement[s] programs that *directly assist New York Businesses in preventing pollution . . . and educat[es] stakeholders on pollution prevention approaches*" (12). Additionally, the agency noted that it held "stakeholder meetings . . . to bring together industry experts, municipalities, business groups, and others . . . to develop short- and long-term strategies and solutions to address [recycling] challenges" (ibid.). Unlike MODNR, NYDEC provides a small amount of space in its reports for discussing the punishment of polluters and climate change; however, the department ensures to point out that efforts like the state's partnership in the Regional Greenhouse Gas Initiative (RGGI) are "supported by an extensive regional stakeholder process that engage[s] the regulated community, environmental nonprofits, consumer and industry advocates, and other stakeholders" (NYDEC 2017, 5). Additionally, they emphasize the agency's commitment to the "cost-effective" reduction of CO_2, while "providing benefits to consumers and the region" (ibid.). Even large settlements for environmental violations appear to be aimed at supporting natural resource goals, as a settlement with Honeywell in 2017 was stated to "result in the implementation of 20 natural resource restoration projects and payment of more than $6 million. The funds, which represent the largest Natural Resources Damages settlement ever secured by DEC, will fund additional environmental and recreational projects" (7). Again, the agency ensures to point out that enforcement actions will have tangible benefits for New York residents.

In addition to stakeholder involvement in pollution control activities, industry partners are active in the state's natural resource management. New York's Partnerships for Regional Invasive Species Management (PRISM) brings "together stakeholders to conduct surveillance, and plan and implement prevention and management programs to control invasive species," and NYDEC helped host a Forestry and Wood Products Summit in 2018 that

brought together academics, state and local government officials, and industry
to "discuss growth challenges and explore opportunities for businesses, including research investments, the potential for greater use of clean technolo-
gies, the importance of growing export markets, and expanded promotion
to bolster the [timber] industry" (NYDEC 2018, 11). Additionally, NYDEC
notes that it works hard to improve its relationships with regulated entities.
In its 2018 report, the department describes "the great strides" it has made
"to improve [its] relationship with commercial fishing communities . . .
ensur[ing] fishermen have as much access as possible, without endangering
public health" (22). In its 2017 report, the agency notes that it embraces new
regulations that "relax or eliminate requirements proven to be burdensome
to the regulated community but having little or no environmental benefit"
(NYDEC 2017, 15). The department states that it made these changes after
"two public comment sessions, five public hearings, more than 25 work-
shops and technical meetings with stakeholders and careful consideration of
thousands of comments" (ibid.). Industry partners are valued as a part of the
regulatory process and appear to be given a number of opportunities for input
and education.

NYDEC's commitment to the negotiated compliance style of regulation
that encourages stakeholder involvement and collaboration can also be seen
through its embrace of management-based approaches to address environmental problems, such as the extensive habitat loss in the Hudson River estuary.
Between the early 1800s and the mid-1900s, the U.S. Army Corps of Engineers worked to deepen the Hudson River for commercial navigation, including the construction of dikes, dredging of the main channel, and attempts to
connect islands (Miller 2013). Unfortunately, while these actions were good
for improving shipping, they "resulted in nearly 4,000 acres of shallow-water
habitat, including the near complete elimination of side channels in the upper
third of the estuary" (6). In addition to these issues, a number of industries
used parts of the Hudson River to provide hydropower for various mills,
reservoirs for irrigation, or drinking water. Unfortunately this infrastructure
"degrade[s] water quality, block[s] fish migrations, and interrupt[s] natural
sediment transport to the estuary (Miller 2013, 7). This degradation of the
Hudson River is problematic for a number of reasons, including the river's
use as a drinking water source and for recreational opportunities, its ecological importance, and its use as a transportation corridor for various goods (5).
Thus, in 2013, the NYDEC developed a plan to improve the health of the river.

This plan is intended to address the issues in the river through "public
and private partnerships" and was developed "through significant community participation" (Miller 2013, 2). The plan requires the participation of
a number of stakeholders, including state and federal agencies, localities,

non-profits, and industry. Together, these groups are asked to work toward the implementation of a plan, with the report importantly noting that keeping the river healthy has tangible benefits, including the restoration of fisheries.

NYDEC embraces the involvement of stakeholders in regulatory processes and emphasizes education and collaboration as it engages with regulated entities prior to and during the regulatory process. Additionally, the department seeks to emphasize that its approaches to regulation consider cost and provide direct (and often economic) benefits to New York citizens. However, recent controversies over fracking and other environmental issues have revealed that citizens and former agency employees fear that industry may be given too much preference during the regulatory process.

NYDEC Regulatory Controversies and Industry Preference

New York's controversial ban of fracking in 2014 was preceded by years of arguing between the gas and power industry, regulators, state legislators and governors, and environmental interest groups. Although the NYDEC eventually released the findings of a seven-year review (along with the New York Department of Health's review) that supported the prohibition of high-volume hydraulic fracturing in New York state (NYDEC 2015), environmental groups accused the agency of working closely with the natural gas industry to soften regulation by giving them "exclusive access to draft regulations for shale gas drilling as early as six weeks before they were made public" (Environmental Working Group 2012). The weakening of the regulations, according to the Environmental Working Group (EWG) (2012), resulted in stormwater permits that did "not propose penalties for drilling companies whose runoff shows radioactivity at a level of concern for public health." This permitting process—without penalties—was referred to as "toothless" by the EWG, "tilt[ed] in the favor of industry," and provided "greater access" and "more opportunities to shape the state's drilling plan" than citizens and other interest groups.

In addition to this claim, a former NYDEC biologist accused the agency in 2016 of "engag[ing] in conflicts of interest when reviewing wind projects" (*Watertown Daily* 2016). The former employee stated that NYDEC and the wind industry shared a "close, collaborative partnership rather than an arm's length relationship" and that the "two may have some similar, shared goals." This relationship, according to the *Watertown Daily*, compromised "NYS-DEC's advocacy role in protecting valuable land and wildlife resources . . . in its fervor for promoting renewable energy, particularly wind" (2016). The former employee also accused the agency of letting Upstate NY Power (one of the wind industries working with the agency) write conditions as part of their own permit, suggesting what the employee referred to as a "close, unusual relationship" between regulator and regulated entity.

This accusation—that NYDEC was soft on regulation with particular industries—was made again in 2017, in relation to permits for concentrated animal feeding operations (CAFOs) (Waterkeeper Alliance 2017). A number of environmental interest groups, including a regional chapter of the Sierra Club, an attorney from Earthjustice, and Waterkeeper Alliance, served the NYDEC with a lawsuit that claimed the agency issued permits for CAFO "that lack the basic enforceable restrictions that communities need to protect their water supplies." The groups accused the agency of "leav[ing] it up to the facilities to set the terms of the permit"—a similar claim as those made regarding the agency's involvement with the wind industry.

NYDEC has been accused of working too closely with industry and providing them with an exorbitant amount of influence over the final version of regulations, particularly as it relates to the sharing of plans and penalization. Citizens and environmental groups have also expressed distress at the agency for delaying the regulatory process. For example, earlier this year, Chemical Watch (2019) issued a statement describing the state's delay of enforcement of cleaning products disclosure, and in 2018, NYDEC was accused of intentionally delaying the release of a report that detailed health information regarding a Niagara area sanitation site (Haight 2018). Citizens living near the Gilboa Dam have charged the agency with being a "fox guarding the hen house" and that problems with the dam are due to a "failure to conduct proper, thorough inspections" of the dam (Dam Concerned Citizens, Inc. 2009). In all of the complaints I have detailed thus far, citizens asked for more input into the regulatory process, some claiming that they had been ignored in the favor of industry. Thus, although NYDEC has been active in helping to ban fracking in the state and has regularly denied permits to polluters (see, for example, the controversial denial of the Competitive Power Ventures permit in 2018), citizens, environmental groups, and former employees still express that the agency works too closely with industry during the regulatory process.

All of this being said, it is not unusual for environmental agencies to be accused of having "cozy" relationships with regulated entities. It is imperative for regulators to work closely with those that they regulate, as they rely upon industries for the accurate disclosure of information. "The effective use of government power depends on information about the conditions of the world, strategies for improving those conditions, and the consequences associated with deploying different strategies" (Parson, Coglianese, and Zeckhauser 2004, 277). The effectiveness of environmental regulations is dependent upon agencies like MODNR and NYDEC having accurate information regarding the ability of industry to cope with particular requirements, the extent of pollution, and the industry's current technology in use. Therefore, regulators must maintain working relationships with industry that enable efficient and effective regulatory analysis (Parson, Coglianese, and Zeckhauser 2004; Sappington and Stiglitz 1987). Given this, we would be hard-pressed to find an environmental agency in the United States that has not been accused of being "too cozy" with the industries they regulate (see allegations regarding PHEPs and mini-EPAs in chapters 4 and 5). However, the allegations I outline in this section illustrate an important point about the complications that may arise from placing environmental regulatory programs within agency cultures that value partnerships and negotiation with local industry: citizens and environmental groups are quickly put at odds with state agencies when flexibility and cooperation with industry appear to take favor over regulatory standards.

From these two cases, it appears as if the early conservation backgrounds of both agencies continue to shape agency preferences and priorities today. In addition to valuing the continued protection and management of state resources, both agencies value collaborative and cooperative relationships with industry partners during the regulatory process—something environmental groups and citizens have not failed to notice. In fact, citizens in both cases have claimed that this flexibility with industry has led to lax regulations or the improper or delayed implementation of regulations in place. However, these two cases speak only to the two states analyzed. Thus, in the discussion that follows, I evaluate my propositions more broadly by speaking with agency employees of NREPS about their goals, priorities, and experiences in working in a combined agency. From these interviews, I garner further evidence that supports the proposition that NREPs value a flexible regulatory process and that that flexibility often leads to more lenient regulatory outputs than those preferred by the EPA.

THE EFFECT OF THE NREP DESIGN ON AGENCY WORKERS

From 2014 to 2019, I spent over twenty-five hours interviewing employees of environmental agencies across the states, including those who work in NREPs, PHEPs, and mini-EPAs. In these interviews, I queried compliance inspectors, legislative liaisons, administrators, epidemiologists, engineers, and others about the goals of their respective agencies, the allocation of resources and attention among their agencies' numerous tasks, and their experience or feelings related to the combination of environmental protection with areas such as public health or conservation. In the discussion that follows, I analyze the conversations I had with NREP employees, in regard to their agencies' approach to regulation and enforcement, how that approach might differ from that of the EPA, and how these employees felt about their multi-mandate agencies.

As is the case with agency reports from NREPs, NREP employees often highlight the cooperative and flexible nature of their agencies, speaking of how their agencies' primary focus is on "helping to get industries back into compliance," rather than punishing them into compliance. One agency employee stated that their agency "interacts a lot with outside stakeholder groups," with "more flexibility to entertain new ideas, [to] think a little bit differently about a project." They want to "help companies find more cost-effective ways to stay in compliance and save money . . . to ensure regulation works, but also [that the agency is] helping [facilities] to do it in a cost-effective fashion." Another NREP employee emphasized that "management plans [are] really the most important," along with "a lot of recognition for industry" when they are doing the right thing. This employee stated that they "try to get industry positive press for their sustainability . . . publish[ing] articles once every couple of months in various media." In all of these approaches, working closely with industry is key to what these employees see as successful regulation. Rather than issuing punitive measures, the idea is to assist, educate, and encourage industries into compliance through the use of market incentives or planning processes. However, this means that environmental protection may be more lenient. For example, one employee noted the following about his agency's regulatory preferences:

> the first thing that gets supported is economic development. Environmental is a secondary consideration, and especially with regulatory programs, there is push back by the entities that are regulated and have a lot of say on different boards/commissions. A lot of pushback, general feeling among staff and certainly something you sense from higher level managers. We don't want to rock the boat. Work with the entities; focus on compliance assistance rather than penalization. Resources aren't there [for regulation]. There's not really a strong regulatory emphasis. It's not even spoken about; people just believe it's there—it's just naturally there. The way people think it should be.

That being said, most employees—even those who expressed distress about the state of their regulatory programs—expressed that cultivating relationships with industry was an important part of their job. However, they were also quick to note that these relationships with industry often put them at odds with the federal EPA and its priorities. The stress from balancing local economic pressures and needing to show the EPA that they were holding industry to federal environmental standards was clear in each interview I performed.

According to an NREP employee, the EPA relies on tools such as inspections and fines, but that their agency "[does not] want to just levy [a] fine." This employee argues that their state-level agency has done a "good job with being more innovative," and that their efforts are "a bit better . . . than EPA." Another employee agreed, stating the following about the EPA's approach to regulation:

> there is a "huge difference between EPA's approach and the approach of [an NREP agency]. EPA is extremely focused on enforcement, endless bureaucracy. Everything has got to be done in certain ways, following different procedures, worried about accountability. Way too strict. I say that being [the EPA's] tendency toward being strict with enforcement. They will shut down everything and not even allow progress to take place. The state is not as strict. Certainly not focused on enforcement very much, but they do a lot of good things with state parks, trails, conservation programs. EPA would levy a maximum fine and make an example out of someone, and at the state-level they would work with the entity and negotiate a fine—if there was a fine at all."

Both employees note that their agencies are more flexible with industry than the EPA is willing to entertain. And, yet, as another interviewee points out, much of agency "funding is dictated by the EPA." Because of these differences in priorities and regulatory preferences, the agency does not "always have the money to do what [they] want." Instead, the agency must rely on its partnerships with stakeholders, including industry. This leads to closer relationships between regulator and regulated entity, as is described in the following:

> We are probably more in touch with stakeholders than the EPA is because we are in the trenches, and we are interacting with local industry. . . . We are much closer to the regulated facilities and the other local organizations and government. It's easier to get into conversations with those groups because we have ongoing relationships. We do a lot more outreach, education, even though we don't have adequate resources. We know the playing field, the players, and how to get something going.

Almost every NREP employee I interviewed described wanting more flexibility from the EPA in how they deal with industry—industry that these state-level agencies feel as if they are more familiar with and, thus, better-equipped to handle. The employees also expressed that they disagreed with the EPA's "strict" approach to enforcement and that there is "sometimes disagreement about which tools [are appropriate] and what [the agency should] prioritize."

These disagreements highlight one of the primary differences between NREPs and the EPA—the preference for negotiated compliance versus enforced compliance. For NREPs, as I describe in chapter 2, stakeholder-inclusive approaches to regulation that involve collaborative and cooperative relationships between the regulated and regulators are an important part of how agencies consider alternatives, such as whether to address a problem using command and control or market incentives or whether to assign a monetary penalty when an industry violates environmental rules. NREP employees see a clear difference in their regulatory preferences from those of the EPA.

In asking employees about the design of their agencies and how that might affect the agency's goals and priorities, many noted that the separation was certainly apparent in day-to-day work. For example, when asked about which goal was the agency's top priority, interviewees often responded that it depended upon the division of the agency in question; goals were not necessarily universal across the agency—even the top goal. One employee stated that even though the divisions are under one department "there is a very distinct separation between the two . . . the funding is completely different. Types of people are different. Really, for the most part, [the divisions] are joined only in name." One issue that helps deepen the divide between environmental enforcement and conservation is that regulatory programs tend to be more controversial. Those that worked in voluntary programs or conservation programs that dealt little with regulation mentioned their good fortune. As one employee noted, "other programs, more popular, more voluntary programs, [the agency] emphasizes those—they try to downplay the negativity of enforcement." This means those programs receive more resources. A number of employees agreed with the sentiment that "the funding for the regulatory program[s] is not sufficient, fees are not sufficient, [and] staff are not sufficient." One employee even noted that due to the lack of resources in regulation, every time they tried to perform their job, "there's push back" from upper management. This "emphasis and greater support for more popular programs" leads agencies "to emphasize things that are not regulatory in nature because they are more popular," and according to one employee, this occurs, "even when causing widespread pollution."

However, even given what employees have argued are poorly staffed, funded, and executed enforcement programs, most of my conversations did not end with employees firmly against a continued combination of conservation and environmental enforcement. One employee stated that while separation might mean "managers would know that [enforcement] was their sole job and they would do much more [of it]," the continued coordination between divisions was important enough to keep them close. In fact, many note that a separation may mean more power for the EPA and, subsequently, less freedom in how they choose to address regulation. While regulators noted their frustration with not being able to execute strict enforcement when necessary, they also expressed opposition to the EPA's typical approach to dealing with industry. These NREP employees want the freedom to regulate the way they see fit; however, they also want the financial and political support to enforce the regulations they have carefully crafted. Unfortunately, the combination of conservation with environmental protection does not appear to provide them with this balance; rather, it appears to create deeper divisions and leave employees with one choice or another. Generally, the employees felt powerless about these organizational decisions that appear to have a significant impact on their work. One employee remarked that when the last organizational change was made, they were "not sure why they did it"; it was "way above [their] paygrade." They told me, "I think I just got an e-mail."

As I have noted throughout the book thus far, conservation is a foundational part of American environmental politics. The conservation movement and the laws and bureaucratic formations that followed helped build toward the momentum that eventually led to the creation of the EPA and a national promise to control pollution. However, for some states, the practice of conservation, including a preference for negotiated compliance and a focus on stakeholder inclusivity, more directly influences the day-to-day regulatory activities of pollution control agencies. For NREPs, the direct combination of the more collaborative and cooperative conservation approaches with environmental enforcement appears to encourage regulators to take on more flexibility with the industries they regulate. And, although the EPA continues to incorporate more industry-inclusive regulatory tools, this focus on flexibility by NREPs does still appear to create divisions between the EPA and its subnational counterparts. Additionally, even the *perception* of a cozy relationship between industry and regulators leads citizens to question the intent of environmental regulators and to question their own safety.

In the chapter that follows, I move on to the PHEP agency design, where public health becomes central. Here, we might expect there to be less fear over agencies that risk the public's health to protect the economy. However, even the combination of two such closely linked areas as environmental protection and public health, creates divisions and confusion over agencies' priorities and preferred approach to dealing with industry. As with the NREPs we have evaluated here, the shifting of focus to a different—even if similar—mandate motivates regulators to focus on those activities that are most politically, socially, and economically rewarding. The organizational design and the regulatory style and approach it encourages and insulates, continues to remain a powerful force.

NOTES

1. Although resource extraction may be managed by conservation agencies in some states, it is important to note that it is often dealt with in agencies labeled as "energy" departments/boards, as well.

2. See the appendix interview methodology and questionnaire.

3. Missouri state parks have remained a part of the Department of Natural Resources, Missouri's environmental protection agency, even with the advent of the Department of Conservation.

Chapter Four

Environmental Enforcement in Combined Public Health and Environmental Protection Agencies

In the nineteenth century, a time that Winslow (1923) refers to as "the great sanitary awakening," human understanding surrounding public health advanced to focus on how social and environmental conditions could be improved to protect the public from disease. However, this was also a time of industrialization, which meant increased pollution and population density that made the roots of disease difficult to assess and control. To address these growing issues, early sanitary surveys were conducted in states, such as Massachusetts and New York, and public health departments and boards were established in New York City, Louisiana, California, Maryland, Virginia, and Minnesota (Institute of Medicine 1988). According to the Institute of Medicine (1988), "by the end of the 19th century, 40 states and several local areas had established health departments." Initially, the control of pollution—both air and water—were housed in these local and state-level public health departments; however, the focus had been on the "problems of overpopulation in confined spaces; to the communicable diseases related to poor housing, lack of water, and sewer infrastructure; and to unprotected food supplies . . . there was little attention to the impact on the environment of uncontrolled and unregulated economic development" (Kotchian 1997, 249). Starting in the mid-twentieth century, the focus shifted to the control of industry, made official by the establishment of the EPA and the passage of numerous pieces of landmark environmental legislation. When states were asked to create plans to implement new federal law, many of their existing pollution control programs were located in public health departments or boards or in conservation departments. For most states, the creation of their own EPAs involved combining pieces of those pollution control programs from public health departments to create a single-mandate agency tasked with setting and enforcing environmental standards. However, a number of states simply invested more control in their public health agencies, assigning them a new regulatory role.

As of 2018, only five states have continued to keep that structure—combining public health and environmental enforcement into a single agency with multiple mandates.[1] As we know from the previous chapters, these states are absent any geographic clustering, even more so than those agencies that adopted the NREP structure. And, these states differ significantly in physical environment, ideology, and the types of industry or natural resource extraction upon which their economies rely. What these states share in common is that their apparatus for protecting the environment is tasked with a number of other health-related tasks—a combination of mandates that spans even more tasks than those implemented by NREPs. For example, these agencies can handle areas such as community health initiatives, disease prevention and food sanitation, emergency preparedness and response, medical/ recreational marijuana regulation, and the administration of Medicaid (one of the largest state programs by budget and number of enrolled citizens) and health insurance exchanges. As with NREPs, this combination of mandates is likely to lead to some tasks receiving more priorities than others, and as Kotchian (1997) points out, environmental health has historically received less priority over other public health issues when environmental health is handled by public health officials. This is likely the case when those environmental health activities take on the regulation of industry, where consideration of cost and stakeholder inclusion—two important parts of public health programming—may not be possible or recommended by the EPA. Additionally, the localized nature of public health programming, where county public health departments are funded and act locally, puts public health officials even closer to citizens who may be economically affected by regulatory decisions. Thus, even when two areas as intertwined as public health and environmental protection are combined, regulation may look different than it does for mini-EPAs.

I examine this proposition through the analysis of case studies of PHEPs and interviews of PHEP employees. In this chapter, I look closely at the Kansas Department of Health and Environment (KDHE) and the Colorado Department of Public Health and Environment (CDPHE). Both Kansas and Colorado's environmental enforcement efforts have long been housed with other public health initiatives; however, the states differ politically, geographically, and in regard to the importance of natural resources, agriculture, and industry. Through these cases and a number of interviews with PHEP employees I find further evidence that the combination of environmental enforcement with an additional mandate leads to more flexible and industry-inclusive enforcement.

CASE 1: KANSAS DEPARTMENT
OF HEALTH AND ENVIRONMENT

The state of Kansas was "an early national leader in state health programs," starting with the establishment of its State Board of Health in 1885 (Shepherd et al. 1999, 12). The origins of environmental health functions within Kansas's health programs can be traced back to the Bureau of Sanitation, which served as one of the major divisions of the State Board of Health, focusing primarily on drinking water quality, waste water, and solid waste management in the 1950s. As Shepherd et al. (1999) points out, the Bureau of Sanitation was loosely related to the University of Kansas's School of Engineering, as the focus of the Board's activities were considered to be "engineering-based disciplines" (13). In fact, the Board's staff even worked in the basement of KU's School of Engineering. It is the Bureau of Sanitation that would eventually become the Division of Environment within the modern-day KDHE.

Following the creation of the EPA in 1970, Kansas considered creating an independent environmental agency; however, "it was ultimately decided that environmental activities in Kansas would remain overseen by the Board of Health" (14). There was great reluctance on the part of many public health officials to remove environmental programs from the purview of public health boards, in which human health was the top priority. In 1974, Kansas took on a large-scale reorganization of its state government, creating a cabinet system that included a new Kansas Department of Health and Environment. Although there were arguments for a separate pollution-control agency that included concerns for the secondary role environmental health played in public health agencies, environmental regulation programs would be implemented as a part of the public health program. As Shepherd et al. argues, it was the "recognition of the inextricable relationship between health and environment" that motivated Kansas's choice to place new regulatory programs under the control of the state's public health bureaucracy.

Over the years, state officials have challenged the combined public health and environmental protection structure of KDHE on at least three separate occasions (with ongoing debates between 1992–1993, constituting one occasion). Arguments have centered primarily around the lack of focus placed on environmental protection programs over the past four decades. Proponents of a separate agency have argued that environmental programs receive less resources and attention within public health and that public health officials are not equipped to inspect and monitor industry activity (Shepherd et al. 1999). This lack of capacity and a focus on public health programs and priorities may lead to more lenient approaches to environmental regulation and enforcement.

The Influential Public Health Framework in the KDHE

The federal expansion of the environmental agenda from 1960–1970 "to include recreational values of the environment; sustaining and protecting wildlife . . . protecting endangered species; protecting unique ecosystems; and aesthetic values" emphasized an already growing mismatch between the goals of public health officials and environmental regulators (28; Harkins and Baggs 1987). In Kansas, this distance has grown even further, as KDHE has taken the lead on implementing health care finance programs (i.e., managing the Affordable Care Act within Kansas). Indeed, environmental protection appears to now be only a small part of the programming implemented by the KDHE.

In KDHE's 2011–2012 Annual Reports, KDHE speaks about public health programs and issues significantly more than environmental regulation (Hopper 2013). The primary focus of KDHE, as related by annual reports, appears to be disease control, through immunization programs; educating the public; health surveillance; and sanitation efforts. This preference for public health (and now, also, health finance) programs is also reflected in KDHE's spending. Environmental expenses make up less than 3 percent of KDHE's (2013) total budget.[2] It is clear from KDHE's annual reports and budget that environmental protection is only one of many tasks within a large and diverse agency. In addition to a primary focus on public health programs and initiatives, KDHE explicitly emphasizes the consideration of cost and inclusion of stakeholders (specifically industry) in the determination and enforcement of environmental regulations.

In its 2015 Annual Report, KDHE states that their Bureau of Air works "to prevent the negative impacts of intrusive federal government regulations" (4). This occurs through "cooperation with businesses to ensure a careful balance of protecting the public health and natural resources of the state, while striving to minimally impact public and private sector business functions" (8). This cooperation with industry includes a dedication toward providing assistance with developing strategies and "helping business understand regulations which apply to their industry." For example, KDHE states that they prioritize keeping permit turnaround times as short as they possibly can and by working closely with industry partners to "keep them up to date on impending regulations" (4). Although KDHE does reference regulatory actions in strategic frameworks and annual reports, regulatory actions are routinely justified as helping to make sure that there is "an even playing field for new businesses to start up" and that "the risk of starting a new business venture becomes lower" (KDHE 2013, 11; KDHE 2015, 10). They seek to achieve compliance "at minimal cost" (11) and claim to have "facilitated an increase of business activity in the State with new construction projects and business expansion at current industrial sites" (7). A primary goal of regulatory activities, according to KDHE, is to work with industry so that industry can continue to expand (KDHE Annual Report 2015, 4). Aside from focusing on economic development, KDHE emphasizes that any required reduction in air pollutants is specific to those pollutants that "impact public health" (7). For KDHE, regulatory actions are crafted specifically to help educate businesses into compliance and to help industry to continue to profit. Any more burdensome regulations are emphasized to be aimed at environmental problems with established human health risks. However, generally, KDHE claims to avoid "costly" regulations altogether.

As with NREPs' collaborative adaptive management approaches, we can also look at KDHE's work with industry to see reflections of the public health prioritization of stakeholders and consideration of cost. KDHE's Watershed Restoration and Protection Strategy (WRAPS) is described as a "process [that] offers a framework that engages citizens and other stakeholders in a teamwork environment aimed at protecting and restoring Kansas watersheds" (Kansas WRAPS 2011). In partnership with Kansas natural resource agencies, KDHE implemented WRAPS to ensure that citizens and stakeholders are able to provide input. In this strategy, industry is able to have a seat at the table. The regulators are in partnership with regulated entities. This approach is very similar to the collaborative adaptive management approach used by NREPs and conservators, in which partnership with local industry is valued.

In addition to KDHE's apparent preference for including industry stakeholders in the regulatory process, the agency may struggle to address environmental protection issues due to the traditional structure of public health programs. In a 1999 report on the organization of public health and environmental functions in Kansas, Shepherd et al. note two structural issues that make the implementation of environmental regulation difficult for public health officials. First, Shepherd et al. (1999) argue that public health officials are trained to focus on populations at risk, as opposed to geographical areas at risk. Due to the nature of environmental issues, focus must be on geographic areas (e.g., a hydrologic unit, an area of groundwater, or a specific ecosystem); thus, an approach aimed at populations at risk would be ill-equipped to manage environmental issues. Additionally, as I note previously, the funding mechanisms for public health and environmental programs differ importantly, with environmental programs relying heavily on federal agencies for funding and technical assistance—funding and resources that may differ significantly depending upon the administration in control. Conversely, public health programs tend to be more reliably funded by state and local tax bases, which provide public health officials with more flexibility in how they choose to spend funds (Shepherd et al. 1999, 30). Also, this reliance on local taxes makes public health officials more susceptible to pressure from citizens and state governmental officials. All of this and more are cited by Shepherd et al. (1999) as potential reasons for ineffective or inefficient environmental programs in Kansas (i.e., the argument in favor of separating public health from environment). Thus, it is not surprising that citizens and state officials periodically question the continued combination of public health and environment in Kansas. In particular, one recent scandal has placed the department into the national spotlight, raising doubts about the department's dedication to strong enforcement.

KDHE and the Sunflower Electric Power Corporation Scandal

In 2006, the Sunflower Electric Power Corporation announced that it would build three additional coal plants to accompany an existing coal plant in Holcomb, Kansas. Although the proposal for a permit was eventually amended to only include one new plant and approved by KDHE permitters, KDHE secretary Rodney Bremby unilaterally denied the permit in 2007, stating that he had concerns over greenhouse gas emissions levels. After many years of then-Kansas governor Kathleen Sebelius vetoing bills that would have allowed the plant to be built, upon the entrance of her successor, Governor Mark Parkinson, the permit was allowed for reconsideration, and Bremby was removed in 2010. In December of 2010, KDHE decided to issue the permit. In words that sound eerily similar to the allegations made against NREPs in chapter 3, the *Kansas City Star* reported that "Sunflower officials and KDHE regulators had a cozy relationship and Sunflower was writing many of the responses to public comments that KDHE received from individuals and organizations. KDHE used the Sunflower responses as its own" (Dillon 2013). Environmental groups, like Earthjustice, accused KDHE of "essentially hand[ing] over its duties to regulate air pollution to Sunflower during the permitting process." The *Star* also reported that KDHE employees worked overtime, even on the weekends, to make sure that the permit would be approved before new federal air regulations that would potentially stop the project would be implemented in January 2011. It is important to note that agency workers were working toward the approval of the permit, even before Bremby was replaced in 2010. Bremby—after being removed—claimed that the permitting process for Sunflower was "not a benign, pristine, routine bureaucratic process. Unfortunately there were abuses" (Metz 2011).

Although the EPA has also been accused in this situation as not providing appropriate oversight of the permitting process in Kansas, in 2011, EPA officials claimed that KDHE "incorrectly informed the [Kansas Supreme] Court" about the EPA's assessment of the plant's potential health risks and that KDHE did not comply with the Clean Air Act (Metz 2011). According to the head of the EPA's Region 7, Karl Brooks, upon discovering this missing or incorrect information, EPA would now scrutinize not just the permit but the "whole decision-making process that produced a permit" (Nelson 2010). KDHE officials denied wrongdoing and pointed to the economic benefits of pushing the coal plant through the process. The permit was "negotiated" for economic benefit, and health risks were established to be within an acceptable range.

For KDHE, this controversial permit approval has resulted in suspicion surrounding the department's commitment to environmental protection. However, it is clear from the agency's documents that they are driven toward helping industry to prosper and protecting economic interests. Additionally, with fewer and less reliable resources to work with than the public health part of the agency, working alongside industry to ensure compliance may be a less costly choice—both politically and economically—than aggressive regulatory actions. Thus, the department's efforts to work with Sunflower to help pursue a permit are not necessarily extraordinary. Also, given the more conservative political climate in Kansas, it is not surprising that efforts to adjust market forces were met with challenges at every turn. The conservative nature of Kansas politics could be the primary motivator of this regulatory behavior. However, a close look at Colorado's PHEP suggests that politics may not be the only factor worth considering.

Case 2: Colorado Department of Public Health and Environment

Physical environment and natural resources have played a significant factor in the cultural and economic identity of the state of Colorado (Satterlee 2017). As American settlers poured into the Western frontier seeking economic opportunity and mass fortune, Colorado became a prime destination due to its extensive "bounty of energy resource wealth" (1). However, this sudden influx of settlers spurred environmental problems, from pollution to natural resource scarcity—problems that Colorado has faced repeatedly during the "booms" and "busts" that occur with methods of natural resource extraction such as fracking. Unfortunately, the citizens of nineteenth-century Colorado had little recourse for addressing natural resource exploitation. Legally, the mechanisms simply did not exist.

As the conservation movement swept across the United States, Coloradans struggled to balance their enchantment with the physical beauty of the land with the economic opportunities brought by continued and unmitigated natural resource extraction (Satterlee 2017). It was not until the environmental movement of the mid-twentieth century that Colorado really began to address the degradation caused by sustained extraction. As Satterlee (2017) describes, actions by the state's governor and attorneys general helped to ensure that tax structures and other incentives motivated more sustainable natural resource extraction activity. The development of boards, coalitions, and state departments helped ensure the enforcement of these new laws and programs.

Also during the late nineteenth century, Colorado began to deal with growing public health issues, such as tuberculosis and diphtheria, along with general sanitation problems that spurred the growth and spread of disease. To address these issues, the State Board of Health was established in 1877. In 1893, Colorado legislators also invested power in local boards of health. The State Board of Health managed the collection of vital statistics and the creation and enforcement of public health regulations until 1947 when the Department of Public Health was formed. This transitioned the State Board of Health to an "advisory, consultative and judiciary branch," while transferring implementation duties to the newly created department ("Colorado's Public Health System," 7). In 1994, the department changed its name to the Colorado Department of Public Health and Environment, where there exists the "combination of public (human) and environmental health in one state agency" (7).

As is suggested by the separate paths of development of conservation and public health policy in Colorado, there are a number of different boards, commissions, task forces, and departments that manage environmental issues in Colorado. However, the primary enforcer of pollution control programs (i.e., the enforcement of EPA regulations under the CAA, CWA, etc.) is the CDPHE. The CDPHE is a massive government agency with 1,380 employees working in one of nine major program areas, including health; environment; marijuana; birth, death and other vital records; public records; laboratory services; health equity; emergency preparedness and response; and lead (CPDHE "Services and Information" 2019). In addition to the agency, there are a number of commissions and boards that carry out adjudication when pollution laws are violated, approve funding, establish fees, and certify facilities (e.g., the Air Quality Control Commission, Solid and Hazardous Waste Commission, Water and Wastewater Facility Operators Certification Board, etc.). These boards and commissions play a significant role in rulemaking and enforcement.

Although Colorado has made a significant amount of progress toward cleaner air and water, the natural resource extraction issues that plagued the state for well over a century continue to be problematic—as does the pollution that has followed massive population growth in the state. Even so, the state's environmental programs appear to be secondary to public health programs. This lack of prioritization, along with the power of external stakeholders and the localized nature of environmental programs in the state, reveal how the PHEP structure adopted by the state may be affecting enforcement behavior.

The Prioritization of Public Health and Public Health Approaches in CDPHE

According to CDPHE, the agency is focused on providing "high-quality, cost-effective public health and environmental protection services that promote healthy people and healthy places" (CDPHE "Strategic Plan" 2016, 3). The agency seeks to accomplish this through "evidence-based" practices from both the public health and environmental fields. That being said, the agency appears to prioritize both its public health programs and a public health approach to environmental issues within the agency. In CDPHE's (2016) strategic plan, they introduce a number of initiatives that they plan to focus on over the next half decade; however, only three out of the twelve initiatives the agency flags as priorities deal with environmental protection. The rest include a focus on obesity, mental health/substance abuse, marijuana, and health care access. In 2011, Colorado laid out a list of "winnable battles" related to health and environment. Healthier air and water were two of these battles; however, in the state's most recent "plan for improving public health and the environment," these are not highlighted issues. Instead, the state refers to these as "other winnable battles" and relegates them to the end of the report with less page space.

This lack of goal priority appears to transfer to a lack of action, as well. As of FY 2018–2019, environmental programs, including air pollution control, water quality control, hazardous materials and waste management, and environmental health and sustainability make up only 15 percent of CDPHE's total budget (according to a Joint Budget Committee Hearing in 2018). This lack of budget prioritization leads the state to fall behind, in regard to environmental spending per capita. According to the Environmental Council of States (2017), the states spend around $47 environmental dollars per person (with a median of around $30). However, Colorado spends less than half that amount at about $15 per person (averaged from 2010–2017). Therefore, while spending only a small portion of a full public health budget on environmental programs may not always mean a state is spending less on the environment than average, in Colorado, the small slice of the budget dedicated to environmental protection does mean fewer dollars spent on environmental protection in comparison to other states. And this continues to be the case, even as Colorado has faced growing issues with increased pollution from oil and gas extraction and population growth.

In addition to the focus on public health goals, initiatives, and spending, CDPHE also approaches environmental issues, using the tools and approaches preferred by public health officials, including prioritizing the consideration of cost, stakeholder involvement, and local implementation of policy. In regard to cost, CDPHE regularly refers to "cost-effective" programs and services and underlines how the cost of regulation is worth the economic benefits provided by the continued pristine environment. For example, CDPHE stated in a strategic plan that "Colorado's healthy air, clear streams and other natural resources are critical to the state's economy and identity" (CDPHE, "Shaping a State of Health" 2015, 25). Specifically, the agency points to air pollution as economically problematic due to the degradation of "visibility at national parks, wilderness areas and other scenic vistas in Colorado vital to tourism, recreation, and the quality of life" (ibid.). When turning to recycling, CDPHE points out that recycling has a number of tangible benefits, including "green jobs" and the ability to conserve natural resources for future generations. In nearly every discussion regarding pollution control as part of its strategic plans/vision, CDPHE includes economic well-being as an important motivator of regulatory behavior. The use of tools like cost-effectiveness analysis to determine the trade-offs and economic effects associated with a policy is widely utilized by public health officials. As Gross, Teutsch, and Haddix (2007) point out "these analyses are important because public health resources are limited and difficult choices must be made. In a world with limited resources, effective interventions that provide the most benefits relative to costs could be expected to be favored" (366). However, as I note in chapter 2, the consideration of cost in environmental protection is explicitly discouraged by some federal laws, such as the CAA, and may not be considered as extensively. Thus, CDPHE's preference for cost consideration may lead the state to take a different approach to environmental protection and enforcement than the EPA.

Another way public health agencies seek to control costs is through stakeholder involvement in decision making. For CDPHE this often happens through the direct appointment of industry, academics, and others to the boards/commissions that advise and adjudicate pollution control policy in the state. For example, CDPHE's Solid and Hazardous Waste Commission ("charged with promulgating and adopting rules pertaining to solid and hazardous waste; establishing fees, issuing interpretive rules and appeals of administrative law judge determinations regarding administrative penalties for hazardous waste violations") is made up of nine members, including three members from industry, three members from the public, and three members from either local government or academia (CDPHE, "About the Solid and Hazardous Waste Commission" 2019). While these commissions directly involve industry in the consideration of enforcement and rulemaking, industry involvement is something that CDPHE points to as an important part of encouraging compliance. For example, the Regional Air Quality Council (RAQC) states that they have been able to work with "external stakeholders" to create new programs, such as an Emission Reduction Credit program, that "incentivize sources to shut down or modify high emitting sources to permanently reduce emissions" (Colorado Air Quality Control Commission 2017, 19). RAQC also mentions that they have begun to inform facilities of air conditions so that they may voluntarily reduce emissions during days of high pollution. Voluntary programs, market incentives, and the ability of stakeholders to influence the enforcement process relax some of the constraints on industry and likely improve relationships between regulators and regulated entities.

Lastly, CDPHE has historically and continues to implement a significant number of public health programs and initiatives via local public health agencies and boards. Public health nursing in Colorado in the late 1880s and early 1900s helped to motivate the creation of a number of local public health departments in the early to mid-1900s. By 2008, local health departments were serving "24 counties and 85% of the state's population" and "were responsible for the provision of a broad scope of public health services within their jurisdiction" ("Colorado's Public Health System" 2019, 9). However, some citizens of Colorado still lacked access to public health services. To address this, the state passed the Colorado Public Health Act in 2008, which required the development of five-year state and local public health improvement plans (based on the use of assessments) and the engagement of communities in health improvement initiatives. Accordingly, a number of CDPHE's programs and initiatives are carried out via local public health agencies (overseen by local public health boards). For example, in the required five-year state plan for health and environment, CDPHE lists both state and local strategies for each policy goal, including improving air and water quality. As an illustration of what this might look like, in improving air quality CDPHE notes that localities determine some of their own regulations to address "concerns related to fireplaces, woodstoves, restaurant grills, open burning, fugitive dust, asbestos, mold, carbon monoxide, and vehicle idling" (CDPHE 2015, 28). Along with crafting some of their own regulations, local health agencies and boards are also involved with educating the public about air and water quality issues and helping to determine what solutions are appropriate for particular communities. All of this local policy implementation is in addition to the presence that localities are often required to have on the commissions and boards involved in adjudication of environmental enforcement.

This fragmented public health structure is not unusual by any means; however, there are a few reasons why this type of implementation structure may be problematic for environmental protection. Primarily, communities and counties will be most susceptible to the economic effects of regulation if, say, an industry is heavily penalized or forced to close its doors. These communities are likely to be working closely with industry to help incentivize continued economic improvements. Thus, local public health regulations and the implementation of those regulations may not be as aggressive as if performed by the state or EPA. Even for regulations not implemented directly by local health agencies/boards, the local presence on state-level environmental commissions helps to ensure that community preferences are prioritized. Additionally, it is unlikely that smaller communities or counties made up of smaller communities will be able to pull together the expertise necessary to adequately create and implement both health and environmental policies. Thus, the quality of those policies and their implementation likely differs across the state. This is certainly the case for controversial oil and gas issues that have recently plagued Colorado, where local initiatives and power differ.

Colorado's Natural Resource Extraction Controversies

According to a 2019 report issued by the Colorado Oil and Gas Association, Colorado's oil and gas sector "employed about 30,000 people in 2017, created 51,000 additional jobs, [and] added about $13.5 billion to Colorado's domestic product" (Staver 2019). This report was compiled in response to a promise for new and stricter regulations on the oil and gas industries by legislators and Colorado governor, Jared Polis. The push for new regulations follows an oil and gas boom that has left environmental interest groups and some citizens distraught over what they have called a "flawed monitoring, inspection, and enforcement framework that has allowed the rapid increase of oil and gas development over the past two decades to continue fouling Colorado's air" (Glick 2019a). According to Glick (2019a), the number of "active wells" in Colorado doubled from about 22,500 in 2002 to 53,102 in 2019. This surge in gas and oil activities is concerning to some because oil and gas extraction releases chemicals during multiple stages of the extraction process that can cause neurological damage, damage to the liver, pulmonary damage, and irritation to skin and the respiratory system, among other effects (Garcia-Gonzales et al. 2019). Additionally, citizens were outraged and frightened by an April 2017 home explosion that killed two and injured two others. The explosion was linked to an abandoned and severed gas line connected to an oil and gas well near the home. In response, citizens have pointed to the CDPHE and the Colorado Oil and Gas Conservation Commission (COGCC) as being too lax with regulation.

A specific complaint citizens have made about oil and gas regulations deals with an exemption for oil and gas companies that allows for those companies to operate for ninety days before applying for air permits that would then set limits on their emissions (Finley 2019). In a 2019 *Denver Post* article, Finley argues that while oil and gas companies are expected to install technology to minimize emissions during that ninety-day period, state inspectors are not checking to ensure that the technology is installed or is operating effectively. Finley (2019) states that this rule has resulted in the release of around three tons of "volatile organic chemical pollution" a day in communities along Colorado's northern Front Range (5). Some have taken these accusations a step further stating that "the state does not comprehensively monitor or regulate the [oil and gas] industry's air pollution emissions. In fact, records show that oil and gas companies are still releasing unknown levels of smog-generating and climate-altering pollutants that are not being measured or reported" (Glick 2019a, 1). Why is this happening? According to citizens, environmental groups, and former CDPHE employees, there is a culture at CDPHE and other commissions that oversee oil and gas extraction that prioritizes industry interests.

Control over oil and gas extraction in Colorado is governed by two primary bureaucracies: CDPHE and COGCC (technically housed within Colorado's Department of Natural Resources). As with a number of natural resource extraction bureaucracies, COGCC is an agency that is tasked with a double mandate: to both foster the growth of the oil and gas industry while also regulating the industry. Citizens and environmental groups point to these competing mandates as problematic and leading to an inappropriate relationship between regulators and polluters. In particular, environmental groups have pointed to COGCC's lack of permit denials as evidence of this inappropriate relationship. According to Conservation Colorado, COGCC "has a nearly uninterrupted 68-year history of failing to deny permits for oil and gas companies to drill—regardless of the risks that wells pose to health, safety, and environment" (Wheeler 2019, 3). To address this, Colorado passed a bill in 2019 that serves three purposes: (1) to change the mandate of the COGCC to prioritize public health, safety, and welfare; (2) help give local communities more control over oil and gas activity; and (3) to encourage regulators to review, tighten, and create rules to further reduce emissions. However, this change does not directly address CDPHE, where environmental groups and former agency employees see another problematic relationship between industry and government.

In 2019, a program director for WildEarth Guardians stated that state agencies like CDPHE work so closely with industry during the regulatory process that the system is "rife for abuse" (Glick 2019a, 4). In particular, the program director, Jeremy Nichols, stated that "the state will say that they're working cooperatively with industry to reduce emissions and that these violations are minor" but that the state is too reliant on "industry–self monitoring and reporting" (ibid.). According to Glick (2019a), attempts from either COGCC or CDPHE to confirm the data provided to them by industry are challenging, due to a "chronic underfunding." CODHE department director Jill Ryan agrees with this assessment, stating that the "Air Pollution Inspection Division, in particular, 'has been under-resourced for years, and maybe even decades'" (5). Furthermore, a former inspector who worked as an air quality inspector for CDPHE for twelve years stated that the agency has only nine inspectors to oversee around fifty-three thousand wells in the state (Glick 2019b). Even with new changes at COGCC, this former employee—Jeremy Murtaugh—told the *Colorado Independent* that the gas industry "wields disproportionate power" over CDPHE, who prioritizes "negotiating with operators instead of enforcing the law and levying meaningful penalties" (Glick 2019b, 2). According to Murtaugh, there is a "mismatch" between Colorado officials' boasting about having some of the nation's strictest regulations on the oil and gas industry and worsening air quality in Colorado. He attributes this growing pollution to CDPHE's "friendliness with industry" and the subsequent "lack of enforcement" (Finley 2019, 8). Murtaugh argues that "'nobody's really looking' to track the amounts—or the impacts—of different toxic air pollutants and climate altering gases," due to the complexity of the extraction process, the lack of agency resources, and the agency's pro-industry culture (Glick 2019b, 2).

Glick (2019a) points to a particular incident that supports Murtaugh's complaints about a "pro-industry" tendency in CDPHE. According to Glick (2019a), Extraction Oil and Gas, an energy company based in Denver, violated state permitting requirements by producing "more than 300,000 barrels above the production limit in its [permit] application and vastly exceeded its VOC emission limits" (12). While the Air Pollution Control Division did notify Extraction of these violations, the agency agreed to an early settlement (around $51,450) and also granted Extraction a new permit that allowed it to produce more barrels and release double the amount of VOC emissions (Glick 2019a). According to Glick's (2019a) interview with former CDPHE employee, Jeremy Murtaugh, Extraction pumped "about $15 million worth of extra oil and gas" addressed by this settlement (12).

In response to allegations like these, CDPHE's Air Pollution Control Division director has noted the difficulty the division faces in regulating each facility adequately. Glick (2019a) describes an interview, in which the director—Gary Kaufman—agreed that a lack of resources, the number of wells in Colorado, and the variety in the sources of pollution makes enforcement difficult. However, Kaufman also points out that the state has made progress in addressing pollutants associated with oil and gas extraction. The department head of CDPHE, Jill Ryan, agrees with Kaufman that resources are stretched thin; however, she also notes that the "culture" of CDPHE "may need to change" as well (Glick 2019a, 12). She states in an interview with Glick (2019a) that her early conversations with CDPHE employees suggested that some regulators may have felt "oppressed in wanting to do more" (ibid.).

In general, the state acknowledges the importance of its relationships with industry but also acknowledges the challenges the agency faces in addressing obstacles as overwhelming in scope as the most recent Colorado oil and gas boom.

As I note in chapter 3, the accusations of industry preference aimed at both KDHE and CDPHE are not unusual for environmental agencies, given that these agencies must work closely with industry to secure accurate and complete information about pollution. However, the degree to which agencies engage industry determines how much influence industry may have in the process and, subsequently, how much they are able to pollute. These cases suggest that stakeholder involvement is an important component of environmental protection for both KDHE and CDPHE. Both agencies are dedicated to working alongside industry partners (along with local governments, experts, and others) to develop regulations and to negotiate enforcement of those regulations. Additionally, both agencies appear to be concerned with the costs imposed by regulation, which is not surprising given that the public health structure of governance places many environmental protection activities in the hands of local officials—those closest to the potential economic impacts of regulation. Lastly, for both agencies, public health issues appear to take preference over environmental issues. Even though there is significant overlap between environmental protection and public health initiatives, this overlap does not appear to help highlight regulatory programs. As was the case with NREPs, interviews with PHEP employees offer further insight into PHEP agency culture and how the combined structure may help solidify particular norms and preferences.

THE PHEP DESIGN, ACCORDING TO PHEP EMPLOYEES

Interviews with PHEP workers provide detailed accounts of the regulatory values embraced by PHEPs, how those regulatory values may differ and compete with the preferences of the EPA, and employees' opinions on the continued combination of public health and environmental enforcement activities. As is the case with the interviews performed with NREP employees, the interviews I detail below were performed between 2014–2019 with a number of PHEP officials of varying positions and ranks from both the environmental regulation and public health divisions of the agencies. These interviews provide further support for the findings garnered from the KDHE and CDPHE case studies: public health programs are prioritized in PHEPs, and environmental regulation in PHEPs tends to be more cooperative and collaborative with industry than regulation executed by the EPA. While interviewees disagreed about whether splitting public health and environmental protection into separate agencies is the appropriate solution to underfunded and lenient regulatory programs, most employees admitted that the current combination or a different combination—such as creating an NREP—would continue to weaken environmental enforcement programs.

The interview script that I have used with environmental agency employees over the past five years asks some preliminary questions about the goals and priorities of employees' respective agencies. One of these questions asks specifically, "What is your agency's top priority?" In my interviews with NREP workers, interviewees often listed a number of agency activities, really unable to put a finger to a so-called "top" priority. This was not my experience with PHEP workers. Almost every worker I interviewed did not hesitate to state that public health was the top priority of the agency, and this did not differ by the agency division in which the interviewee was employed. One difference between division workers, though, was that public health officials often felt that this prioritization was appropriate, making statements, such as "the whole reason for protecting the environment is to protect public health." Conversely, environmental health or environmental protection employees stated that the assumption that environmental protection was completely about public health was somewhat misleading. For example, one employee stated that while employees on the public health side of the agency sought to aim regulation only at those problems that directly affected human health, environmental regulators were concerned with a much broader, long-term picture. The employee argued that environmental employees "want *all* streams and water clean, even if people are not around." The employee continued, stating that "there will be health people and environmental people [working together] and to some extent we don't really speak the same language. Our missions go separate ways, and they may overlap, and they may not sometimes." Another environmental PHEP employee agreed, expressing that "the environment is less prioritized [than public health]." This regulator described the irony of the word "environment" even being part of the agency's name, when the agency was clearly more dedicated to the "health side." For these employees, the lack of focus on environmental program activities and successes hurt environmental regulators' reputation with the public and industry. The regulators felt that their work was not taken seriously. According to other PHEP employees, this lack of prioritization has tangible consequences.

In discussing the effect that the PHEP combination has on environmental programs, an employee flatly stated that they "get little or no assistance or enhancing of [their] goals by the agency, as a whole." Others agreed, particularly in regard to staff and monetary support, asserting that environmental programs "don't have enough resources to do anything [they] are supposed to do" and "are all understaffed." As one employee noted, "everyone is short on money." Additionally, employees spoke about the preferred regulatory approach of their agencies and how they were encouraged to work closely with industry. One staff member stated that "basically everything is . . . talking to stakeholders," while another employee noted that the task most important to their everyday work was to work closely with stakeholders, particularly since this particular employee's program was funded directly by stakeholder permit fees. Thus, to keep stakeholders happy and engaged, the program works to "invite [stakeholders] to events . . . to educate and work together." The goal of this collaboration, as another employee noted, is to help industry to correct their own behaviors, and many PHEP employees "strive for voluntary compliance in everything [they] do," rather than having to force compliance through penalization. For many employees the following sentiment resonated: "if every day you tell someone to do something they don't have the funding for or could lead to penalties and violations, it can be really defeating." Thus, as a regulator articulates, PHEP employees "don't drag [industry] into court and make them comply" or "chase" industry. Working with industry makes enforcement more comfortable for both regulator and regulated entity.

For some employees, this focus on keeping industry content (and, thus, more willing to cooperate), is discouraging. One interviewee had moved from a regulatory position in another state and described their regulatory experience in the PHEP as much different. They stated that they "have to slow down and have to say, 'let me help you mitigate the problem,' even though [they] are used to saying 'you aren't following the regulation, and you need to.'" This staff member described a "different mind-set" in the PHEP, where "public health will work with [industry] and no consequences." Others described how the close relationships with industry helped to downplay certain environmental problems and lead the agency to focus primarily on those problems that could directly affect the economy. For example, one worker described how the agency was determined to focus on cleaning up the state's lakes, even though the EPA had encouraged the agency to focus their attention elsewhere. Why focus on the lakes? The employee argued that the agency saw those recreational spaces as economically important; the cleanup of those lakes has tangible monetary benefits without burdening industry.

This kind of economic focus puts PHEP at odds with the EPA, as employees note that their state-level agencies "may be less likely to enforce," due to a lack of power over industry. One employee described the stark differences between the EPA's preference for regulation and that adopted by their PHEP, arguing that the PHEP was "more interested in trying to find compliance assistance with regulated entities than EPA would be" and that the PHEP provided "more outreach, more technical assistance, help with

compliance, letters after letters of warning, rather than large penalties."
According to this worker, their PHEP was generally "more gentle with the
regulated" than the EPA. Another worker agreed, claiming that the focus of
their environmental programs was "not as much on fines and penalizations"
and that "the way [the PHEP] does fines is much more reasonable," while
"the EPA will generally slap on gigantic fines." The interviewee concluded
by stating that "the state regulatory attitude is not always in sync with the
EPA regulatory attitude"; the states "have a much better context with local industry." However, this difference in regulatory preference proves
problematic due to state environmental agencies' reliance on the EPA for
funds. Many employees lamented over this dependency and how the EPA's
funding "determined what [they] focus on and spend money on." For these
workers, "it's easier to do what the state wants to do. The feds have different
mandates," but, often, state employees feel that they "have to do things that
don't apply to the state," but have to be done if the agency wants to continue
to pull in funding. Some employees even describe how they sought to avoid
some of this pressure by turning to industry to help fund their regulatory
programs through fees. This allows PHEP workers to "purposely avoid having to pursue [EPA] goals."

The difference in regulatory preference between the EPA and PHEPs is
clear, and it appears to drive a wedge between federal and state bureaucrats;
however, even those bureaucrats most frustrated with the EPA's push for
strengthening enforcement see some of the challenges inherent with the
PHEP combination. One of these employees stated that "if there were two
different departments, it would help with support and funding," since public
health is currently more prioritized. Another employee agrees, arguing that
they "would prefer a separate agency because of the focus and prioritization."
While the "concept [of the combination] is wonderful," the promised integration between public health and environmental protection has failed to come
to fruition. Thus, environmental protection suffers from less attention and
fewer resources without the connection to public health that could potentially
improve environmental protection efforts. Unfortunately, PHEP employees
noted that "there isn't a lot of that connectivity" between environmental
regulators and public health officials. Environmental regulators feel as if their
programs are simply a line-item in a massive budget, where the priorities of
the agency do not allow them to focus solely on the environment. Even those
in favor of continued combination note that "logistically combinations are
challenging" and that a separation may lead the agency to become "more in
line with the EPA's ideas." Almost all of the PHEP employees described past
conversations between lawmakers, citizens, and bureaucrats about the potential separation or a new combination (e.g., with agriculture, natural resources,

etc.). These conversations revolved around what employees described as a general suspicion surrounding the PHEP structure—that it led to weaker environmental protection.

The combination of public health and environmental protection feels like an ideal combination, given the meaningful overlap between environment and health. There is a reason why many of the earliest pollution control activities in the United States were planned and executed by local health departments. However, the broad meaning of "public health" encompasses so many mandates that environmental programs are often only a small portion of what PHEPs implement. As these case studies and interviews have reflected, this lack of visibility leads to less support and fewer opportunities for increasing funding and resources for enforcement. And, importantly, the combination also encourages regulators to work more closely with industry stakeholders and to consider cost—a preference that is held by public health officials but is also reinforced by the localized nature of public health policy implementation and funding. Thus, it is likely that in states with PHEPs, industry has more power over the regulatory process, and environmental enforcement reflects a preference for negotiation rather than forced compliance. Even when environmental protection programs are combined with programs as complementary as those found in public health agencies, the combination changes the way states implement environmental enforcement.

NOTES

1. As of the publication of this book, North Dakota's Environmental Health Section of their Department of Health became an independent agency—the North Dakota Department of Environmental Quality (NDDEQ).

2. It is important to note that KDHE has taken control over health finance, which now makes up just short of 90 percent of KDHE's expenditures.

Chapter Five

Environmental Enforcement in Mini-EPAs

The organization of the U.S. EPA could have taken a number of different forms, including the creation of a federal NREP, combining natural resource management with environmental protection, the creation of a federal PHEP, enveloping environmental enforcement into the already existing Department of Health, Education, and Welfare, or the creation of a small agency with power to set standards within the Executive Office of the President. However, the President's Advisory Council on Executive Organization (1970) determined that anti-pollution programs "be merged into an EPA, a new independent agency of the Executive Branch," an agency which would serve as the "principal instrument" for repairing environmental damage and establishing standards to prevent further degradation. Advisors of President Nixon wrote that the current government was "neither structured nor oriented" to effectively implement environmental protection and that the disjointed nature of anti-pollution control programs, which separated air, water, and land programs, created a lack of coordination and a failure to understand the environment as a complex, unified, interrelated system (Ash Memo 1970). Additionally, continuing to house anti-pollution programs in agencies with other primary mandates was distracting those programs from environmental protection goals. In order for the environment to be a priority, the president's advisors believed that pollution control programs needed to be housed independently with a sole focus on preventing and addressing environmental problems. Nixon took this advice, and the EPA was structured as those on the advisory council recommended.

For many states, the structure of the EPA served as a blueprint for their own environmental agencies. For some states, the adoption of a mini-EPA structure quickly followed the mandate for state implementation plans put forth by the passage of policies, such as CAA and CWA; however, for others, this structure was adopted later, as demand for a single-focus agency grew. Currently, thirty states structure their environmental agencies in a model similar to the EPA. For most agencies this means that the state's executive branch houses an agency that manages air, land, and water by setting environmental standards that abide by federal regulations and by enforcing those standards through inspections and adjudication. States differ in their use of various boards and commissions to oversee this process; how they deal with other areas of the environment, such as protecting endangered species or disaster response; and how active local governments may be in implementing environmental protection policies, but the mini-EPA agencies all have one thing in common: they are structured solely to protect the environment through anti-pollution regulatory programs. And, like the EPA, these programs rely heavily on a command and control approach to regulation.

The 1970 Ash memorandum did not only spell out suggestions for the EPA's structure; it also included recommendations about the new agency's functions. According to the memo, the EPA should be organized to "observe, record, understand, and predict the atmospheric and aquatic environments"; perform the research necessary to develop sound criteria and new pollution abatement technology; and to "set and enforce standards consistent with national government goals." More simply, the EPA's primary functions were to set and enforce standards that were based on evidence-based best practices. Thus, the EPA was created with the purpose of implementing a command and control approach to environmental protection. And, although the Ash memo does include instructions about considering the economy and working with the private sector as part of the regulatory process, the agency would soon discover that economic considerations and industry involvement would often put them at odds with environmental goals. For environmental policy analysts, the "cost" of clean air was irrelevant, when the benefit was human survival.

The command and control approach adopted by the EPA is based on a market failure theory of environmental regulation, in which "it falls to government to impose standards that force private actors to effectively internalize environmental costs" (Huffman 2000, 26). Command and control policy instruments involve the following: (1) the mandate or prohibition of specific behaviors or the use of technology; (2) the definition of some standard or criteria; (3) and the monitoring of behavior and enforcement of that criteria when behavior is out of line with set standards. Some of the benefits of the command and control approach are that the setting of targets and standards is relatively inexpensive, and the goals are clear; however, the uniformity of regulation is unable to accommodate diverse industries, who may see the cost of compliance as too high when it has not been shaped for their situation personally (Bengtsson et al. 2010). Additionally, monitoring costs can be exorbitant, as the enforcement of standards requires regular inspections. In general, command and control regulation has been labeled by industry as an approach that creates an adversarial relationship between regulators and regulated entities, where costs may soar without the explicit consideration of economic effects. While the EPA does incorporate other regulatory tools, such as voluntary compliance programs and market incentives, the command and control approach still drives the agency's primary day-to-day regulatory activities.

For the states, shaping their own environmental agencies in the mold of the EPA also means that the structure is likely designed to regulate via the command and control approach, an approach that requires the creation of uniform rules/standards and a focus on the enforcement of those standards. While we would expect states to be more flexible with industry than the EPA, given their relative economic sensitivity to industry interests, the additional mandates that may allow for even greater wiggle room in NREPs and PHEPs do not exist for mini-EPAs. Like their federal counterpart, the sole focus on environmental protection allows these agencies to prioritize environmental protection over all other interests and considerations. In mini-EPAs, enforcement should be an agency priority, with less negotiation than we might expect in states where other interests compete and other regulatory approaches are preferred/encouraged.

In assessing this expectation, I evaluate the mini-EPAs of California (California Environmental Protection Agency—CalEPA) and Texas (Texas Commission on Environmental Quality—TCEQ). Economic competitors California and Texas are both populous and geographically large states with abundant resources; however, they are politically quite different, with Texas being more conservative than California—an ideology that generally rejects or seeks to contain the regulatory apparatus inherent to mini-EPAs. Comparing two states on different ends of the ideological spectrum allows us to better understand the influence of environmental agency organizational design. As with the previous case studies, I evaluate CalEPA's and TCEQ's history, development, and current stated goals and activities, along with evaluating the public perception of the two agencies. These cases provide additional evidence for the argument that these organizational decisions matter and that enforcement may take more priority in states where it is placed independently from other mandates. This is likely the case, even though mini-EPAs may face more or less pushback from state governments and citizens that differ significantly in their preferences for regulation and environmental protection programs.

CASE I: CALIFORNIA
ENVIRONMENTAL PROTECTION AGENCY

The state of California is a stunning collection of the some of the country's most beautiful landscapes, which include mountains, deserts, coastline, ancient forests, and the greatest number of national parks in any state. The weather is temperate and creates excellent conditions for growing an abundant amount of produce, wine, and—now legally—marijuana. In addition to its impressive aesthetics and agriculture, California is also host to a number of natural resources, including gold, timber, and oil, among others. Along with being the most populous state in the country, California also has the largest gross domestic product of any American state and is the largest agricultural producer in the United States and the third-largest oil producer. Given its size, economic power, diversity of resources, and breadth of land and regional cultures, California is often referenced as an ideal laboratory for policy experiments and has arguably been a driver of national policy for that reason. This is particularly the case for environmental policy, where the state is regularly referenced as the primary leader on environmental issues—even to the point of coining the term, the "California-effect" (Konisky and Woods 2012a; Vogel 1995; Vogel 2018).

Here are a few examples of California's leadership on environmental policy.[1] Only fourteen years after California officially became a state, the Yosemite Valley became the first publicly protected wilderness area in the United States. About twenty years later, in 1885, California was one of the first states to begin the regulation of logging and to encourage reforestation. By the end of the nineteenth century, three out of the four national parks in the United States were located in California. But, the state's environmental efforts did not stop with conservation. California was the first state to enact its own air pollution control statute in 1947, with cities like Los Angeles performing research on air pollution that would guide the rest of the country in developing their own regulatory rules and programs. During the environmental movement of the 1960s and 1970s, California continued to lead, creating the nation's first emissions standards for motor vehicles, establishing the first coastal protection agency in the United States, and adopting the first energy efficiency standards for appliances and building codes. In more recent history, California has been a leader on climate change, passing some of the most ambitious legislation in North America. However, even as an environmental leader, California continues to face numerous environmental challenges including urban air quality problems, farming inefficiencies, and drought and water quality issues, among others.

California's leadership in environmental protection is interesting for the purposes of examining state environmental agencies because the state's own mini-EPA was not established until 1991. Prior to the establishment of the CalEPA, many of the state's environmental programs existed within the state's Resource Agency or were spread out among other agencies/departments. For thirty years after the establishment of the federal EPA, California debated the creation of a single-focus environmental protection agency; however, there were a number of political officials, interest groups, and industry stakeholders who feared the movement of environmental programs into a single regulatory arm who fought to keep the programs within their original jurisdictions (CalEPA 2001). The only organizational action taken in the "environmental decade" of the 1970s was taken by Governor Edmund Brown, Jr., who was unable to establish a separate agency dedicated to environmental protection but was able to create a cabinet-level position to help advise the governor on environmental issues. This individual would run an informal "environmental affairs agency," which meant that they would be responsible for coordinating all environmental activities and managing resources. Although California continued to make progress on environmental issues during this time, it was clear that the state was suffering from a lack of coordination between its environmental protection programs (CalEPA 2001). Thus, in 1991, CalEPA was established. It currently consists of the following divisions: the Air Resources Board; the Department of Pesticide Regulation; the Department of Resources Recycling and Recovery; the Department of Toxic Substances and Control; the Office of Environmental Health Hazard Assessment; and the State Water Resources Control Board. As with many other states' mini-EPAs, CalEPA's structure closely mirrors the federal EPA's, which contains similar divisions or program areas. CalEPA states that it pursues its mission of protecting the environment by "developing, implementing, and enforcing environmental laws that regulate air, water and soil quality, pesticide use and waste recycling and reduction."

CalEPA's Environmental Enforcement Preferences

California's environmental regulations are often described as some of the "strongest environmental policies ever passed" (Schmidt 2007, 146). And, this is true. The CAA allows California to enforce pollution standards that go beyond the federal standards, and the state has taken advantage of this, passing some of the strictest regulations on pollutants like particulate matter (ibid.). Additionally, California has fought to go beyond federal standards in regard to controlling greenhouse gas emissions by enforcing strict standards on automobile emissions and reaching out to other nations after the Trump administration's withdrawal from the 2016 Paris Climate Agreement. The state's current environmental secretary, Matthew Rodriquez referred to the current federal administration's rollback of environmental efforts as "the latest bouts of lunacy" (CalEPA 2018, 1). In a number of cases, California has gone its own way on environmental protection. However, CalEPA's approach to regulation and enforcement still mirrors the federal EPA's command and control approach. CalEPA regularly details a number of enforcement activities and programs and publicizes the actions taken against industries in violation. Additionally, even when the agency employs other regulatory tools, such as cap-and-trade, strict limitations are placed on industry, and industry is expected to make changes that align with the agency's own risk assessments and pollution research.

In looking at CalEPA's reports and agency documents, the primary difference between this agency and NREPs and PHEPs is clear: enforcement is out in the open. While NREPs and PHEPs speak rarely or quietly about enforcement data, information about large penalties or settlements, or day-to-day enforcement activity, CalEPA proudly advertises a yearly report on enforcement, titled the "Environmental Compliance and Enforcement Report." In this report, the agency details the enforcement programs implemented and actions carried out by each division. The aim of enforcement actions, according to the 2016 report, are to "help ensure achievement of anticipated emissions reductions and assure a level playing field for all regulated entities" (CalEPA 2016, 6). The Air Resources Board states that this is necessary to achieve their goals, as their enforcement actions bring "out-of-compliance companies into compliance with air emission requirements, and assess penalties to deter future non-compliance" (ibid.). The reliance on monitoring, compliance assistance, and enforcement mechanisms such as violation letters, penalties, and settlements is not unique to CalEPA by any means; most states rely on these tools. However, CalEPA's public proclamation of enforcement activity is somewhat unusual, particularly in comparison to those reports from NREPs and PHEPs. Additionally, it is made clear by the report that "vigorous enforcement" is a necessary part of regulatory activities, "complementing" other compliance assistance and outreach programs and helping to "motivate violators to *promptly* return to compliance" (CalEPA 2017, 8, emphasis added). These enforcement reports also contain detailed accounts of industry violations and the punishments that were assigned to these industries to help rectify damage. In 2016, the report contained summarizations of violation activities and punishments assigned to large companies, such as Apple, FedEx, and Shell.

One of the reasons CalEPA is able to articulate all of this information and to execute the numerous enforcement actions they highlight is that the state has invested in environmental enforcement through a grant program that provides funds for training environmental prosecutors and regulators. This program is the result of a 2002 California statute (Penal Code 14300 *et seq.*) that states that "the enforcement of California's environmental laws is essential to protect human health, the environment, and the state's economy" and that "fair and uniform enforcement of laws and regulations governing the environment benefits law abiding businesses, firms, and individuals." Thus, the legislature agrees to fund a "support program that assist[s] local and state enforcement officials in prosecuting violations of environmental laws through the training of peace officers, investigators, firefighters, public prosecutors, and state and local environmental regulators." According to CalEPA, 25 percent of these grant funds are given to the Environmental Circuit Prosecutors Project, California District Attorneys Association, and the California Commission on Peace Officers Standards and Training, respectively. The remaining 25 percent is given to the secretary for other grants related to enforcement, education, and training. Unlike the enforcement and compliance programs of other states, the focus of this program is on helping to expand enforcement efforts, rather than focusing primarily or solely on helping industries back into compliance. In fact, CalEPA defines compliance assistance as encompassing more than outreach, education, and training; it also includes frequent inspections, and the resolving of compliance issues through civil penalties—what many states would label as enforcement activities, rather than compliance assistance (CalEPA 2016).

That being said, CalEPA also prides itself greatly on its use of market incentives, such as the cap-and-trade program the state launched in 2013 to lower greenhouse gas emissions. Applying to large electric power plants, industrial plants, and fuel distributors, the program caps six gases covered by the Kyoto Protocol (CO_2, CH_4, N_2O, HFCs, PFCs, SF_6), along with other fluorinated greenhouse gases. After capping the amount of greenhouse gases, the state auctions emissions allowances during quarterly auctions and provides some free allocations of allowances, depending upon the industry and facilities' relative efficiency. However, the cap is not static. The goal of the program is to lower greenhouse gas emissions by decreasing the cap by 3 percent annually from 2015 through 2020 and by quickening that decline between 2021 and 2030. Thus, industries must eventually rely on emitting fewer greenhouse gases. Although this is a market solution rather than relying on command and control, industries still must abide by the cap in place, along with the auctioning system, or be sanctioned. Additionally, the declining cap means that industry must continue to find ways to lessen their reliance on producing these gases. More simply, California continues to take an approach that places the burden on industry to make changes or face penalties, rather than working to primarily/only craft regulations that make compliance easier.

Industry and Public Perception of CalEPA

While overall public approval of CalEPA is difficult to discern, most public reports surrounding the agency make one clear conclusion: CalEPA is a fierce regulator on behalf of a state that has moved aggressively toward expanding environmental protections. Some argue that this aggressiveness appears to have paid off, citing statistics from the "California Green Innovation Index" (2017) that show that for around twenty-five years, California's GDP and population have grown, while its per capita carbon dioxide emissions have remained static (Tarantola 2018). Additionally, supporters point to the state's job growth since 2006 when the state passed the California Global Warming Solutions Act, job growth which has outpaced the rest of the nation all while carbon dioxide emissions decreased by about 12 percent (Tarantola 2018; "California Green Innovation Index" 2017). As Tarantola (2018) points out, California has remained insistent on strengthening and maintaining environmental protections, even during economic downturns, such as the 2008 recession. In response to heightened economic stressors, the state did not choose to weaken regulations; rather, it chose to pursue further greenhouse gas emissions reductions by implementing a cap-and-trade program. While this program may be more preferable to industry because it allows those polluters who are most dependent upon particular emissions to continue to pollute, a strict cap was still placed on greenhouse gas emissions. Industries were still required to adapt during a time when most states were reluctant to take on new environmental regulations. According to William Fulton, the former planning director for the City of San Diego, this did lead to "blue collar industries" leaving the state, but those industries were eventually replaced by growth in other areas, such as green industry (Ali and Wilkes 2016; Tarantola 2018). Representatives from these industries agree with the conclusion of CalEPA supporters that the agency implements some of the strictest regulations in the country; however, they are not as positive about the impact of those regulations.

During the heat of the Great Recession, local newspapers, such as the *North Bay Business Journal*, published seething critiques of the "notoriously expensive and uncertain regulatory environment" created by CalEPA and California politicians. Then-president of the California Manufacturers and Technology Association (Sacramento), Jack Stewart, argued that continued strict regulation made businesses afraid to invest in California (2010). One example Stewart gives of the invasiveness of CalEPA enforcement was a California company that testified to the state legislature that they "had been inspected by regulators 165 times in 2008, nearly every two days, and that inspections had increased another 26 percent in 2009" (Stewart 2010). Stewart views this type of enforcement behavior as overwhelming and discouraging to businesses thinking of locating in California. Furthermore, Stewart and other industry representatives argue that one primary issue with CalEPA's regulatory approach is that they do not account for the economic impact of their regulations. Industry would like to see an "unbiased, independent economic impact report for every major regulation that's proposed will achieve this" (Stewart 2010). As I have noted previously, it would be uncommon for agencies modeled after the EPA to pursue economic analyses of environmental protections without being required to by law or without additional political pressures. Thus, it is not all that surprising that economic analyses are not prioritized by CalEPA.

Even though industry has continued to push CalEPA toward more industry-friendly regulatory approaches, such as the Greenhouse Gas Emissions Cap and Trade endorsed by the California Chamber of Commerce, industry representatives continue to point to what they perceive as "inflexible" and "over-zealous" regulation and enforcement by CalEPA. Upon a recent CalEPA ban of chlorpyrifos, an organophosphate insecticide used to control soil-borne and foliage insect pests on food and feed crops, the agriculture industry insisted that the ban would "have a chilling effect on agricultural companies looking to invest in California" and that it would slowly kill industries, such as the citrus industry (Hooker and Wyant 2019). CalEPA banned use of the pesticide after finding that "the pesticide causes serious health effects in children and other sensitive populations at lower levels of exposure than previously understood. These effects include impaired brain and neurological development" (Hooker and Wyant 2019; CalEPA, Department of Pesticide Regulation 2018). President of the California Citrus Mutual, Casey Creamer, stated that the findings by CalEPA included "numerous overestimation of the risks and an exaggeration of the actual use [of the pesticide], which wasn't based on sound science" (Hooker and Wyant 2019).

The trucking industry has also expressed frustration with CalEPA over a perceived lack of consideration of cost and economic impact of regulations. The California Air Resources Board has named an ambitious greenhouse gas emissions reduction target of reaching 40 percent below 1990 levels by 2030. Given that the board sees vehicles—particularly those running on diesel fuel—as one of the primary contributors to greenhouse gas pollution, CalEPA has begun to assemble additional regulations on the trucking industry. According to a co-owner of a transport company, these regulations have "been a nightmare" and have forced several small trucking companies out of business (Sparks 2015). The trucking industry states that the regulations are "too costly and difficult to meet"; they—like the citrus industry—would like to see the governor require CalEPA to incorporate financial impact analyses into their rulemaking process. Again, CalEPA is noted as being less flexible and more assertive with enforcement than the NREPs and PHEPs in previous chapters.

Industry and citizen concerns are not only in relation to emissions policies, the control of water in the state is another area in which CalEPA is perceived to be a rigid enforcer. In 2018, the State Water Resources Control Board (under CalEPA) presented a plan to require an average of 40 percent of unimpaired flows on the Stanislaus, Tuolumne, and Merced Rivers to protect salmon. According to the *Union Democrat*, a local newspaper, this would increase flows to the Sacramento-San Joaquin River Delta, while decreasing water flow to the New Melones and Don Pedro reservoirs. Those living in areas with decreased water flow expressed great concern over the potential for fewer water recreation activities, stating that "water recreation drives as much as 80 percent of tourism in the county" (Maclean 2018). Once again, opponents of the policy argued that CalEPA had not "adequately analyze[d] the impacts." A local water policy committee chairman, Karl Rodefer stated to the *Union Democrat* that CalEPA "made the decision of what they wanted to do and wrote the plan, and now they're not listening." Even Democratic California Assembly members, who we might expect to be more sympathetic toward CalEPA, showed exasperation with the plan, accusing CalEPA of violating "principles of good faith."

CalEPA is not only perceived as a zealous regulator by Californians; the agency regularly receives pushback from federal political officials, including even the EPA, itself. Most recently, the Trump administration's EPA attacked Obama-era auto emissions standards as "too high," stating that "the regulations would make it more expensive for automakers to build emissions-compliant cars and that the price would get passed on to consumers" (Geuss 2018). In particular, the EPA's head in 2018, Scott Pruitt, pointed to California's waiver to set its own emissions standards as being problematic. Given California's massive state economy, some of its environmental regulations, like auto emissions regulations, often push industry to make changes across the board, even if other states do not adopt California's standards. Thus, Pruitt stated that the EPA would reexamine California's waiver, as they work to relax auto emissions standards. The chair of the California Air Resources Board, Mary D. Nichols, responded by stating that "California will not weaken its nationally accepted clean-car standards, and automakers will continue to meet those higher standards, bringing better gas mileage and less pollution for everyone" (Geuss 2018). The state has faced federal challenges before (e.g., during the Bush administration); however, it continues to push back and pursue environmental standards that go above national requirements.

CASE II: TEXAS COMMISSION
ON ENVIRONMENTAL QUALITY

The state of Texas encompasses a formidable 268,597 square miles of land, rivers, desert, swamplands, rolling hills, and towering cities. Like California, the state is a top agricultural producer, ranking first in the country in the number of cattle and sheep raised and in the amount of cotton and pecans produced. The state also boasts a thriving fishing industry that relies on one of the largest shrimp catches in the country. Beyond agriculture, the state is immensely successful in the extraction of natural resources, as the state is responsible for about one-fifth of the nation's oil production and close to one-third of the country's natural gas supply. This is in addition to the state's production of other resources, such as lime, salt, and sand. In many ways, the environmental politics of Texas feel like the polar opposite of California. One thing they have in common, though, is the broad collection of land and resources that make the environment an important economic contributor.

Much of this success is due to Texas's natural access to large mineral deposits, fertile soil, and rich grasslands. However, the access to all of this land is something that has also long-attracted businesses and those seeking to benefit from the state's numerous resources. This has often led to confrontations between developers, industry, and conservationists, who have fought fiercely over the state's resources. In response to some of these early fights, Texas began to create conservation policies and boards to oversee those policies as early as the late nineteenth century. In 1895, the state's legislature created the Fish and Oyster Commission to regulate fishing, and the Game Department was included in 1907. A State Parks Board was formed less than twenty years later, as projects of the federal Civilian Conservation Corps helped to add to the state's parklands (Bengston, Blankinship, and Bonds 2003). The Texas Parks and Wildlife Department (a merger of existing conservation boards and commissions) was created in 1963; the department now holds all authority for "managing fish and wildlife resources in all Texas Counties" (ibid.).

Apart from wildlife management, the state progressed in the early twentieth century on other environmental issues. Most early environmental actions dealt with water, including the state's first drainage districts, the creation of the Texas Board of Water Engineers, and the creation of freshwater supply districts. These actions all occurred between 1905 and 1929. The state did not begin enforcing clean water standards until 1945, when the Texas Department of Health was authorized to enforce drinking water standards for public water. These regulations would be followed in seven years by the state's first actions to address air pollution in 1952. These increasing concerns surrounding pollution led to the state's eventual creation of the Texas Water

Pollution Control Advisory Council within the Department of Health, which served as the state's first administrative body dedicated to dealing with pollution. As has been the case in other states we have evaluated, conservation and pollution-control rules, regulations, and administrative bodies developed separately from one another in Texas.

Over the 1960s and 1970s, Texas responded to national stimuli, passing its own water and air pollution control policies and removing its Air Control Board from the Department of Health to make it an independent state agency in 1973. One of the board's first actions was a direct challenge to the EPA in *Texas et al. vs. the U.S. EPA*, in which the board argued that the EPA's failure to approve the state's implementation plan for controlling ozone and the inclusion of additional requirements for Texas's revised ozone plan were a violation of the law. This would be the first of many challenges Texas would aim at the EPA. However, at the same time that this dispute was taking place, the Texas Air Control board completed the implementation of the first continuous-monitoring network in the nation. The state moved forward on environmental regulation, even as some in the state wished to remain in place.

For years, various water, toxic waste, and air programs bounced from one board, commission, or department to another, including the Department of Water Resources, the Texas Water Pollution Board, the Texas Water Commission, the Water Development Board, the Water Rates and Utilities Services Program, and the Public Utility Commission—just to name a few. Given the need to consolidate knowledge and encourage collaboration between pollution control areas, in 1993, the state combined all air, water, and waste programs into a single agency—the Texas Natural Resource Conservation Commission.[2] This transitioned the state into a structure similar to the federal environmental protection agency. One primary difference is that the mini-EPA formed in 1993 holds the same leadership structure as the former Texas Water Commission, with three governor-appointed commissioners that serve a maximum of two, six-year terms. The chair of the commission is chosen by the governor, as well.

Even though Texas has followed an environmental protection trajectory similar to other states, including California, the state is often labeled as an environmental policy "laggard" (see Lester 1995). Popular media headlines include "Texas slashes EPA policies," "How Texas Pollution Cleanup Benchmarks Fail," "Texas Environmental Agency Fails to Address Public Comments on Pollution," "In Texas, Environmental Officials Align with Polluters," and more. This is not surprising, given Texas's reputation for being a conservative, low tax state that is driven to attract and keep as many industries as possible, even if that means fewer regulations. However, this generalization is somewhat misleading. For example, Texas is actually the nation's largest wind power producer, with over 10 percent of state electricity coming from wind (the highest in the nation). According to McCracken (2017), the state is also a national leader in water conservation, using 11 percent lower than the national average of ninety-eight gallons. McCracken (2017) argues that Texas has accomplished this through extensive drought planning, innovation at the local level, and collaboration between local water systems. This, along with help from the EPA's WaterSense program, has kept drought-prone Texas out of some of the trouble that has faced California (McCracken 2017). Additionally, the state is proud to host fourteen national monuments, recreation areas, preserves, historic sites, and national parks—in the top ten for states in the nation.

The story in Texas is more complex than we might imagine. The Texas Commission on Environmental Quality (TCEQ) is directed by law to focus on economic impact of regulations, something explicitly forbidden by federal statutes aimed at the EPA. And, yet, the TCEQ is still a regulatory agency, structured like the EPA. It is not a natural resource agency or a public health agency; it is a mini-EPA. Thus, what does regulation look like in a state where the laws dictate economically sensitive regulation, but the agency is structured to execute implementation like the EPA?

Environmental Enforcement at TCEQ

Perhaps unsurprisingly, TCEQ's framing of enforcement activities in its strategic plans and annual enforcement reports differs from that of CalEPA's. Like many NREPs or PHEPs, there is a substantial focus on helping industry into compliance and ensuring that regulations are necessary and efficient. For example, the agency's mission statement specifies that the agency "strives to protect [the] state's human and natural resources consistent with sustainable economic development. [The agency's] goal is clean air, clean water, and the safe management of waste." Also, the agency focuses on how each of their goals address a set of objectives, including accountability to tax and fee payers, efficiency (minimum waste of taxpayer funds), the use of performance measures to ensure program effectiveness, excellent customer service, and the transparent sharing of information that "can be understood by any Texan" (TCEQ 2016). These goals are pursued by "solicit[ing] input from the general public and regulated entities," promoting "timely authorization and permit processing," and by fostering "voluntary compliance with environmental laws and provid[ing] flexibility in achieving environmental goals." The agency appears to value many of the same values held by PHEPs and NREPs.

That being said, it is clear from agency documents that a number of these goals and regulatory preferences have been required by state law. For example, the set of objectives listed above that includes efficiency and accountability to tax payers is labeled as a "statewide objective." Texas agencies are required to report to their political principals how each of their goals helps to achieve the state's overall governmental objectives. Thus, regulatory enforcement actions—by law—must be justified by showing that enforcement measures are not economically harmful or wasteful. State laws, such as the Texas Environmental, Health, and Safety Audit Privilege Act, which states that regulated entities who initiate a self-audit and voluntarily disclose violations are provided "certain immunities from administrative or civil penalties," directly instruct the agency on how to carry out its regulatory programs (Shaw, Baker, and Covar 2013). Another example is found in Texas House Bill 1794, which limited the extent to which local governments could benefit from fines and penalties imposed on polluters within the local governments' jurisdictions. This directly altered one of TCEQ's options for penalization. Beyond this, Texas statutes cap penalty amounts for particular industry sectors and violations types (Shaw, Baker, and Covar 2013). Thus, the Texas legislature and governor have a significant amount of influence over TCEQ's enforcement actions and general approach to regulation.

However, this does not stop TCEQ from engaging in the regulatory approaches used by the EPA. For example, even when the agency speaks of "promoting compliance" through "voluntary efforts" and by offering incentives, this language is quickly followed by an assurance that these tools are used only "while providing strict, sure, and just enforcement when environmental laws are violated" (TCEQ 2018a, 4). The agency promises "educational outreach and assistance to businesses" but states that they "assure compliance with environmental laws and regulations by taking swift, sure, and just enforcement actions to address violations" (5). Like CalEPA, TCEQ publishes an annual enforcement report, in which they state that their "enforcement orders resulted in almost 247 million pounds of pollutants eliminated, reduced, or the routes of exposure reduced, and an estimated cost of more than $42 million to achieve compliance" (TCEQ 2018b, xi). And, while the agency states that enforcement is "a tool among many" available to the agency, the agency uses the EPA's language to describe the justification for enforcement as ensuring that "violators not come out ahead economically to the disadvantage of those entities that spend substantial resources to comply with the law" (1-4). This holds true, even for some of the state's most important industries, such as petroleum refineries and oil and gas extractors, who made up 25 percent of the agency's administrative and civil judicial orders issued to previous offenders. TCEQ argues that their approach to enforcement is effective, pointing to compliance rates of over 95 percent for most inspected facilities.

In addition to TCEQ's routine inspection and violation discovery process, Texas also employs the Texas Environmental Enforcement Task Force, which TCEQ contends makes the state "a leader and national model in the investigation and prosecution of environmental crime" (3-1). Through a partnership between TCEQ, the Texas Parks and Wildlife Department, the Texas Railroad Commission, and others, this task force works with the EPA's Criminal Investigation Division and other law enforcement and prosecution arms to investigate and help prosecute environmental crimes. Additionally, the task force helps to train local law enforcement in the identification and reporting of environmental crimes. The task force's investigations result in a number of convictions against both corporations and individuals each year, consisting of both felony and misdemeanor counts.

Interestingly, there do appear to be divisional (i.e., air, waste, and water divisions) differences in enforcement behavior at TCEQ that may suggest that some divisions within the agency are more in line with the EPA than others. For example, in FY 2018, water violations made up around 49 percent of administrative and civil judicial orders issued by the agency, and waste violations made up around 32 percent of those orders. Air, conversely, made up around only 13 percent. This could simply be that air pollution is less problematic or better-controlled; however, it is also likely that water availability and quality is more economically consequential, where severe drought, poor drinking water, and polluted freshwater recreation areas are sure to pose consequences. Air pollution may not be as immediately problematic. And, given the overall pressure on state-level bureaucrats to protect (or at least not hurt) the state's economy, it is not all that confounding that TCEQ would focus its enforcement efforts on the most economically problematic types of pollution violations.

TCEQ provides us with a look at an interesting juxtaposition: a conservative state with an EPA-style environmental agency. Given the existing literature on environmental regulation, we might expect that TCEQ would entirely adopt the approach of negotiated compliance, as the agency's conservative political principals push for less stringent enforcement on important economic contributors. And, truly, the agency does emphasize these tools as an important part of their approach to regulation. However, as I note previously, this emphasis is required by state law. And, TCEQ still claims to participate in the kind of strict enforcement of federal and state environmental law that is embraced by the EPA. More simply, any state agency in Texas may be inclined to curb regulation. However, this finding brings to question whether Texas may be even less inclined to participate in traditional command and control regulation and enforced compliance if the state housed a combined environmental protection agency. As the evaluation of the agency's documents suggests, the opinions of industry and Texas citizens that I discuss below also reflect a struggle between state elected and appointed officials who desire fewer and less stringent regulations and an agency that is designed to implement the enforcement of rules and regulations delegated to it by the EPA.

TCEQ: A Disarmed Regulator

Media coverage of the TCEQ almost entirely revolves around one primary complaint: TCEQ is an agency that has been entirely captured by industry and serves the goals of polluters over the health of citizens (for example, Hopkins 2015; Collier 2017; Gregor 2010; Sadasivam 2018). Often, these articles point blame at the agency, itself, stating that it lacks the motivation necessary to monitor polluters and punish those in violation—that the agency has an anti-regulation culture. And, certainly, there is reason to believe that the agency is driven to protect industry in a way that the federal EPA is not. As I note previously, the agency's mission explicitly requires that the agency consider the economic impacts of its regulatory actions. In fact, that requirement is the first part of the agency's mission; clean air and water come second. However, several public evaluations of TCEQ reveal more nuance. The agency is largely constrained by the state legislature and has been rendered so weak that the EPA works through TCEQ to ensure that environmental quality standards are met—something that the mini-EPA structure of the agency helps make possible. The agency, which cannot look to state conservation taxes or public health fees for support, is dependent upon the EPA for funding; therefore, its existence is dependent upon continuing to work with the EPA, even when state officials are fighting to keep the EPA out of Texas.

An example of some of these limitations appears in the events of an April 2013 fire at the West Fertilizer Company in West, Texas. This disaster spurred an ammonium nitrate explosion that practically leveled the small town, killing fifteen and injuring more than 150. Almost immediately following the disaster, local and national news outlets began to point to weak regulations and a lack of coordination between Texas regulatory agencies (and federal agencies) as determinants of the disaster (Henry 2013).[3] Identified weaknesses included a lack of regulation on the storage of ammonium nitrate, a lack of education/training in the handling of the hazardous chemical (particularly during a fire), and the zoning allowances that permitted the building of homes and businesses close to the facility (Henry 2013; Smith 2016). In particular, TCEQ was blamed for negligence after it was revealed that the agency knew in 2006 that the West Fertilizer Company was handling "2,400 tons a year of potentially explosive ammonium nitrate in a warehouse near schools, houses and a nursing home"; the agency had described the area surrounding the facility as "residential and farm land" (Martin 2013).

In response, TCEQ argued that they lacked the statutory authority and training to have properly inspected the facility and to have applied enforcement that would have prevented the explosion (ibid.). And, agency officials were correct: they did not actually have the ability to regulate the storage of the chemicals in the West, Texas explosion or to enforce those regulations with administrative penalties (Henry 2013). In fact, they would not be given that authority until two years after the disaster when Texas HB 942 transferred this regulatory power to TCEQ from the Department of State Health Services. It is often the case that TCEQ lacks the authority or capacity to execute enforcement, even in those situations where the public expects the agency to play an active role. For example, in the case of another explosion that occurred at the Intercontinental Terminals Company in Deer Park, Texas, TCEQ was again criticized for a lack of inaction; however, TCEQ does not hold the "authority to regulate above ground storage tanks" (Thomas 2019). Although journalists, such as Thomas (2019), continue to refer to the "hands-off approach used by TCEQ—and increasingly by the U.S. Environmental Protection Agency," the agency is severely limited by elected Texas officials in its ability to regulate.

Additionally, the powerful Texas attorney general's office and the governor's appointment powers limit or expand TCEQ's powers. In early 2019, Texas attorney general Ken Paxton and Governor Greg Abbott joined forces to file a number of suits against companies for "damage [they] have done to our environment" (Collier 2019). This was much to the surprise of environmental activists, who claim that many potential environmental damage suits that could be taken out against pollution violators have never materialized. Although much of the blame for this is generally aimed at TCEQ, the agency "refers a few dozen civil cases to the attorney general's office every year" (ibid.; TCEQ Enforcement Report 2018). It is entirely up to the attorney general's office whether or not to proceed with those suits. Furthermore, the governor's ability to appoint powerful commissioners to staggered terms that can span as long as twelve years ensures that a particular regulatory agenda can be sought in an agency that is already weakened by law in its enforcement powers.

A further example of these limitations is in the continued winnowing of TCEQ resources by both the Texas state government and the federal government. Even though "Texas is home to the nation's second largest environmental regulatory agency," in 2011, the Texas legislature cut the agency's funding by 30 percent and reduced its workforce by 8 percent (Fehling 2014). TCEQ firmly denied that these cuts altered its capacity, stating that "the agency chose to maintain its core mission of permitting and enforcement in order to retain 'boots on the ground,'" but it is hard to deny that the decreases make day-to-day enforcement more difficult. For some state lawmakers, the agency's funds are entirely debatable, even suggesting that $20 million of TCEQ's budget be moved to an "Alternatives to Abortion Program," that oversees and promotes "crisis pregnancy centers" that discourage abortion (Tuma 2017). The agency does not have stable legislative support.

The state agency has also suffered from cuts to the federal EPA's budget. As the Texas state government continues to slash funding from TCEQ, they become more reliant on the EPA's funding to carry out enforcement programs. Thus, attempts to cut EPA funding are worrisome to TCEQ and other Texas officials. According to Austin (2017), eighteen TCEQ programs receive EPA funds and "if cuts come, they'll likely mean 'fewer cops on the beat' to enforce regulations." Even some state Republicans are concerned about TCEQ's current capacity and its ability to handle additional federal funding cuts (ibid.). TCEQ's dependence on the EPA funding runs deep enough that the agency appears to act as a bellwether for national environmental politics. Under the Obama administration, the weak agency was at the mercy of an EPA that sought to aggressively challenge the Texas legislature's creation of a clean air program that "prioritize[es] the needs and demands of businesses . . . at the expense of public health" (Gregor 2010). Under the Trump administration, however, TCEQ has received less funding and fewer directives about strengthening its regulatory programs. With less support and instruction, the weak agency pursues the path forged for it by its political principals: constrained enforcement.

Interestingly, even with all of these constraints, TCEQ still manages to carry out its enforcement programs, even penalizing some of the state's most powerful industry sectors. For example, as Fehling (2014) notes, even though TCEQ performs "far fewer comprehensive inspections of polluting facilities," compared to other states, TCEQ "took enforcement action against major facilities at about the same rate as the national average and issued penalties in almost all such cases whereas the national average for issuing penalties was 80 percent." Indeed, these penalties are capped by Texas law, but where TCEQ does have discretion—in assigning penalties in the first place—they appear to either meet or exceed national expectations. In fact, these caps exist partially because the agency has—at times—considered strengthening its penalization of violators by instituting enforcement mechanisms, such as mandatory minimum penalties (Metzger 2004). TCEQ appears to do what all state agencies do, which is work within the political constraints of their state.

However, the state does continue to punish polluters, even in the face of pressure from its political principals and industry. For example, in 2018, TCEQ pushed for a change to a backflow prevention code that would help "prevent contaminated water from back flowing into clean streets" (Self-Walbrick 2018). Some stakeholders expressed "vehement" opposition to the rule, stating that it could impose a heavy cost burden on those property owners who would need to buy new equipment (ibid.). In 2019, TCEQ fined a large resin maker—Formosa Plastics Corp—$121,000 for "leaking plastic pellets into waters" surrounding the plant (Toloken 2019). Again, environmental activists expressed distress at the extent of the penalty; however, penalization was carried out. According to TCEQ's yearly enforcement reports, penalizations against large industries are not unusual, with thousands of administrative orders issued each year (TCEQ 2018b).

Texas provides an interesting test case for the proposition that the mini-EPA agency design should result in a regulatory preference for enforced compliance, as the political forces in Texas and the power of industry push against this approach with vigor. While California's environmental progressivism inflates its environmental agency, allowing it to go above and beyond national environmental standards, Texas's opposition to stringent regulation runs directly counter to the style of environmental protection agency it has adopted: an agency whose sole purpose is to carry out environmental regulatory programs. To push against this, elected officials in Texas have pursued as many avenues as possible to shape the agency's behavior. This includes purposing the layout of the state's bureaucratic agencies, where environmental responsibilities are spread across a number of different departments, to constrain the actions for TCEQ. These political and institutional differences between California and Texas are obviously meaningful mediators of how powerful an agency design may be. This is bound to be the case across all of the states.

IMPORTANCE OF COMPARING ACROSS STATES

Variation across the states is one reason that case studies of the various environmental agency design types, alone, are insufficient in determining the nature and plausibility of the theory of environmental agency design. In the preceding chapters, we have explored how agency design may affect the way environmental agency employees think about their goals and values. The combination of natural resource conservation or public health with environmental protection appears to create—at the least—goal ambiguity and—at the worst—goal conflict that cripples some of agencies' most stringent environmental enforcement programs. However, the potential for conflict among the values and goals of combined environmental agencies is likely further shaped by state context. In chapter 1, I discussed the vast literature dedicated to explaining environmental policy in the states. This literature finds strong support for the proposition that a variety of factors unique to each state help drive environmental policy decisions. Political context matters. Economic context matters. The ways in which elected officials and citizens place constraints on their bureaucracies matter. The strength of state institutions matters. "States are different and these differences have a real, direct effect on the lives of their citizens" (Moncrief and Squire 2013, 3).

Thus, in the second part of this book, I compare across state environmental agencies to develop a more complete picture of what factors determine environmental enforcement in the states and how environmental agency design fits in to that picture. First, I test the proposition that environmental agency design promotes a particular regulatory approach by using quantitative content analysis methods to assess agency documents. I, then, construct models of environmental enforcement behavior, evaluating how politics, state economic factors, and agency design affect enforcement behavior, such as the assignment of violations and penalties and the magnitude of penalization. Finally, I evaluate how the differences in capacity of state institutions moderates the influence of elected officials and environmental agency design. From the cases and interviews presented in the first part of the book, it is clear that environmental enforcement behavior is complex; the data and analyses I present in subsequent chapters further reflect that complexity. However, one factor is certain: as is the case in Missouri, New York, Kansas, Colorado, California, Texas, and the states of all those interviewed, the structure, capacity, and preferences of state institutions are consequential to environmental agency goals and values. To consider anything less would surely be a misunderstanding of environmental protection in the United States.

NOTES

1. These examples and others are discussed at length by Vogel (2018).

2. The agency's current title—the Texas Commission on Environmental Quality—was assigned to the agency in 2001 after recommendations from the Texas Sunset Commission.

3. A U.S. Bureau of Alcohol, Tobacco, Firearms and Explosives investigation into the explosion revealed that the cause of the initial fire that sparked the explosion was likely arson.

Part II

COMPARING ACROSS ENVIRONMENTAL AGENCIES

What are the Determinants of Enforcement Behavior?

Chapter Six

The Dominance of Public Health and Conservation in Environmental Agencies

Often in my conversations with bureaucrats, they struggle to describe exactly why their respective agencies prefer particular approaches to regulation—or preferences for approaches to any activity, generally speaking. They mention political pressures, pressure from industry, the preferences of their various supervisors, and the ease of particular approaches versus others. But, at some point, almost each interviewer concludes with a blanket statement—"that's just the way it is." While perhaps unsatisfying, this blanket statement helps capture a concept that Wilson (1989) identifies as agency culture or "the way things are done around here" or "the "patterned and persistent way" of think-ing about tasks (91). Through a combination of traditions, rules, delegation of power, legal constraints, and shared attitudes and beliefs, each organization forms a type of organizational personality (Schein 1990). And, those organi-zational personalities can be powerful motivators and sources of pressure in determining the day-to-day actions of employees.

While specific agency preferences and the reasons behind those prefer-ences are difficult for employees to articulate, it is even more difficult for scholars of organizational theory and public administration to conceptual-ize of organizational cultures. This is problematic because a number of theories across the discipline argue that certain agency characteristics, such as structure, political pressure, employee demographics or professional background, or employee beliefs shape the dynamic of an organization (Gormley et al. 1983; Meier and O'Toole 2006). The theory of environ-mental agency design is one of those theories. Specifically, I argue that the choice to combine natural resource conservation or public health with environmental enforcement has led environmental agencies to develop a sort of "split" personality, where employees must regularly choose between competing preferences. In particular, the preferences introduced by natural resource conservation and public health that involve more cooperative and collaborative approaches with industry, more localized (and, therefore, influenced) control of policy implementation, and more cost-sensitive approaches to regulation, offer state-level environmental regulators with an opportunity to relax some of the pressures placed on them by citizens and political principals (Konisky 2008; Koontz 2002). Before anything else, the theory assumes that organizational design translates into a prioritization of public health and natural resource conservation programs and approaches over environmental protection programs and approaches. Organizational design alters the focus and goals of the agencies, significant components of agencies' organizational cultures.

In the case studies and interviews presented in previous chapters, I provide some support for this assumption, as interviewees from PHEPs and NREPs describe how environmental programs receive less attention and fewer agency resources. Additionally, through agency documents, we are able to see that public health and conservation programs and approaches receive more space and emphasis. However, the small number of states considered through the case studies makes it difficult to know if these differences are systematic (i.e., does this generally occur across combined environmental agencies) or are occurring for other reasons (e.g., politics, economic factors, institutional capacity). Therefore, an important piece of the causal mechanism—that agency design affects the balance of priorities and preferences—remains untested.

DEFINING AGENCY VALUES

The existing research on agency values or preferences has produced a number of compelling definitions and measures. Scholars have calculated measures of agency values and preferences using surveys that ask bureaucrats to rank their priorities and to discuss various agency protocols (Gormley et al. 1983). Others account for the characteristics of bureaucratic workers, theorizing that those characteristics may lead to greater support for some approaches to policy implementation or may help bureaucrats more effectively serve specific clients (Meier and O'Toole 2006). In general, these existing measures of agency values and preferred organizational approaches are used to describe the attitudes and dispositions of bureaucratic employees, which are then considered together to explain something about the agency as a whole. One issue with this approach is that the attitudes of bureaucratic workers, in particular, are not always predictive of agency outputs (Konisky 2008). Another issue is that characteristics of bureaucrats are generally linked to bureaucratic outputs that are likely to be affected by the dynamics of representative bureaucracy (e.g., demographic characteristics, specifically). For example, research has shown that female representation on police forces helped to curb the tendency of police departments in the 1990s to ignore mandatory arrest laws for domestic violence (Chaney and Saltzstein 1998). In that situation, specifically, gender was likely to make a difference; however, for policy decisions such as regulation, characteristics of bureaucrats beyond ideology may not be as likely to impact how environmental agency workers feel about enforcement. Furthermore, because attitudes are not always adequate predictors of bureaucratic decision making, it is difficult to use the template of existing measures to address differences in environmental agency enforcement behavior.

Another component that is often pointed to as being a central component of an agency's culture or organizational personality (those features that are developed out of an expression and adherence to agency values and preferences) is organizational language. The Linguistics Society of America (LSA) (a group of researchers that study language), states that "your culture—the traditions, lifestyle, habits, and so on that you pick up from the people you live and interact with—shapes the way you think, and *also shapes the way you talk* (Birner 2012, emphasis added). Although the LSA says that the direction of causality between language and culture is difficult to dissect, one thing is for certain: language and organizational personalities are related. The LSA is primarily speaking about anthropological culture; however, the relationship surely holds within organizations as well. Take my interviews with combined agency workers. One of the comments made about the difficulty of being in a combined agency was specifically about "not speaking the same language." Every organization has a unique language, and language is reflective of organizational culture (i.e., sentiment). Thus, one valid way to approximate and compare across the values and priorities of environmental agencies, is to compare their language.

SELECTING A SOURCE OF
AGENCY LANGUAGE

In the previous chapters, I rely heavily on agency annual reports, strategic visions, agency website publications, and enforcement reports to help formulate an image of an agency's priorities and regulatory preferences. Unfortunately, this approach presents a number of issues, including that not all agencies publish the same documents or publish the same documents over the same time span. Additionally, many of these reports are done at the behest of a state's governor or its legislature, indicating that the audience is for the agency's political principals. In the discussion of TCEQ, for example, this type of political pressure or the demand for information to be released in a particular way was likely intended to make information reported by the agency slanted in a particular direction (e.g., in that particular case regulation and enforcement was not emphasized). Depending upon the state, the report may be framed more for the use of its employees, more for the use of citizens, or more for the use of legislators and executive staff. Thus, the eclectic collection of agency documents used in my case studies is unlikely to provide a reliable source of comparable agency language across all states.

Another option for official agency language are the rules formulated, approved, and implemented by the agency. These could provide a thorough look into exactly the kind of activities the agency performs. However, even though a rule exists, it does not guarantee the rule will be enforced as it is written, or some rules may leave more or less discretion up to regulators. For example, when environmental inspectors are deciding how to address a particular violation, they may have a significant amount of discretion in determining what their response will be. Simply because the rule indicates that some response must be made, does not confirm what *kind* of response will be made. Additionally, just because a rule exists, does not necessarily mean that it will be enforced *at all*. How do bureaucrats prioritize among the rules they implement? If rules conflict, which rule receives priority? This is likely to be unclear. The process for rulemaking also differs between states, meaning that agencies may have more or less power over the final version of a rule. These differences present difficulties in comparing across multiple states.

Thus, instead of annual reports or rules, which present some of the challenges I have noted above, I look to agency communications with the public, or press releases, as a sample of agency language that encapsulates how agencies prioritize programs and approaches. Admittedly, press releases have not traditionally been used to measure the values of bureaucratic agencies; however, agency communications to the public help agencies to develop a public persona—the way they want to be perceived by the public and their political principals. Because of this, press releases provide valuable information about agency prioritization. Research on the use of press releases supports this, finding that the tone and kind of language used in press releases helps organizations communicate their expectations and opinions about programs/policies or potential policy outcomes and that press releases help agencies to inform outside interests that certain policy goals are being pursued (Davis et al. 2012; Lemov 1968).[1] Thus, press releases are valuable because they can tell us what the agency has done, what the agency plans to accomplish, and what the most important priorities of the agency are—those activities and programs that they would want the public to know about. Furthermore, press releases are public proclamations about agency activity, and because of their publicness, execution within the programs and policies mentioned is even more likely, as the public is able to hold the bureaucracy accountable when they are provided with information. If an agency does not intend on executing a mandate, then, it is unlikely they would bring the public's attention to that program.

From a methodological standpoint, press releases are an advantageous data source, as well. Almost all agencies, at the state level and the federal level, release some kind of notices to the press about agency activities, programs, and successes. Since the massive growth of government services that followed the Great Depression, government has had to inform the public of their goals and how existing programs were meant to achieve those goals (Saunders 1937). Thus, it is easy to compare across many different types of agencies, using widely available press releases.

All of this being said, there are drawbacks to using press releases, as well. As is the case with the rulemaking process, the process through which information travels through a particular bureaucratic agency and the process for determining what kind of information should be shared is likely to be different across the states. Additionally, information could be used selectively to speak to political actors, even if the law does not require that those releases be created for legislatures or governors, specifically. However, even given these limitations, the availability of this data and the relative similarity in the type of information shared makes press releases useful in assessing agency values and priorities.

BUILDING A MEASURE OF AGENCY VALUES/PRIORITIES

There are a number of ways to analyze language. In previous chapters, I searched through agency documents and interviews with environmental bureaucrats to uncover broad themes and ideas that either provided support for or challenged my expectations about environmental agencies' regulatory preferences. Although this kind of in-depth qualitative analysis of language through interviews and summations of archived documents is useful in pointing to general themes in language, there are a number of benefits to using more quantitative approaches to analyzing language, as well. To begin, quantitative analysis of language is "a non-obtrusive, non-reactive measurement technique. The messages are separate and apart from communicators and receivers" (Riffe et al. 2005, 38). There is less room for error in interpretation of messages. Additionally, bureaucratic agencies may be "unwilling or unable to be examined directly" (38). During my recruitment for interviews, it was difficult for me know whether or not the individuals that chose to speak made that decision because of some commonality across their experiences (e.g., unhappiness with their current job/position, familiarity with arguments for/against agency combination, etc.) A quantitative language analysis of agency documents is not dependent upon the selective participation of the agency staff. Thus, analyzing language quantitatively is a way to approximate agency values and priorities, while lowering the risk of misinterpretation and increasing our access to data that would be far more difficult to obtain through surveys and interviews. Admittedly, there are weaknesses to the quantitative analysis of text, including the issues of comparing language across time—words change in meaning—and an inability to pick up on deeply latent messages that are conveyed by tone and body language. However, for the purposes of this analysis—to establish an agency-wide measure of values and priorities, a quantitative analysis of text over the same period of time for each agency should be a valid methodological approach.

In order to develop a measure of agency values and priorities, we first must develop a measure that allows us to compare agencies to some standard for how much a mandate should be focused. More simply, this means that we need to understand what an "ideal" representation of environmental protection, public health, or conservation might look like. Once we have established this, we can then compare environmental agencies to one another to determine how closely they fit this ideal prioritization of mandates and their associated programs. One methodological tool that allows for us to set ideal points and determine how our other observations compare to those points is Laver et al.'s (2003) Wordscores text similarity analysis package for extracting policy positions from political texts.[2] Although the Wordscores method was initially created in order to determine the ideological positions of political parties, the same concept can easily be applied to placing many kinds of organizations on a variety of dimensions, since it is simply a way of using word frequencies to determine how closely language in a set of documents mirrors reference texts that have been placed *a priori* on a dimension (i.e., our "ideal" documents). More simply, if we can establish a dimension and have confidence in the position of a certain number of texts on that dimension, we can also determine the position of other texts. Specifically, the Wordscores method uses the "relative frequencies" that are observed for each of the "different words" in reference texts—those texts that we already have determined a policy position for—to calculate "the probability that we are reading a particular word" (Laver et al. 2003, 313).

For example, in a simple liberal to conservative dimension that ranges from −1.0 to 1.0, the text of a knowingly conservative political party may be a 0.90. Given the policy dimension placement that we have given *a priori* to that reference text, we can "generate a numerical score for each word" that appears in the document (313). The score reflects the expected policy position of the text, "given only that we are reading the single word in question" (313). These wordscores are then applied to the words in "virgin texts"— those texts that we have no *a priori* policy knowledge about—and can be used to estimate the positions of virgin texts on the policy dimension that we are interested in. "Each word scored in a virgin text gives us a small amount of information about which of the reference texts the virgin text most closely resembles," each scored word adding more information about the virgin text and making us more confident in the final score of the text, as a whole (313). In conclusion, we should be able to calculate a score for each text that tells us how closely the language of the virgin text resembles that of the reference texts and how much error is associated with our calculations.

Constructing a Dimension

My use of the Wordscores package requires thinking about our "dimension" in a slightly different way. Rather than a simple liberal to conservative dimension, our dimension represents the ideal prioritization of environmental protection programs and approaches, from the least ideal to the most ideal. For this to work, we must first select documents that portray the most and least ideal prioritization of environmental protection. For our purposes, the federal EPA should act as an adequate comparison point because it is an agency in which environmental protection is prioritized to the greatest degree. Unlike many state agencies, it does not perform additional mandates, and it is the agency to which all other state environmental agencies must be responsive. In determining the least ideal prioritization of environmental protection programs and approaches, a number of options exist, including any agency that does not prioritize environmental protection at all. However, we want to determine whether there are significant differences between how programs and approaches are prioritized between mandates that do often overlap with environmental protection (i.e., public health and conservation). Thus, we need to use an agency comparison point that incorporates the programs and approaches embraced by public health and conservation agencies. For this, I use the Department of Interior (DOI) and the Department of Health and Human Services (HHS) as comparison points for NREPs and PHEPs, respectively. For DOI, this "ideal" classification makes sense as the proclaimed mission of the DOI is to conserve and manage the nation's natural resources "for the benefit and enjoyment of American people" (U.S. DOI 2019). To accomplish this, the DOI focuses on a number of conservation-related programs, including the promotion of energy security, access to outdoor recreational opportunities, and improvement of species and their habitats (U.S. DOI 2019). For DHHS, the classification works, as the agency's primary role is to educate the public on health risks, manage disease, and regulate the health industry (U.S. HHS 2016). Environmental health is related to these programs but not necessarily fundamental. DOI and DHHS represent agencies for which public health and conservation programs and approaches are central.

Upon the selection of our comparison points, we have now established a dimension in which the ideal prioritization of environmental protection programs places the EPA press releases at one end of our dimension, and DHHS and DOI press releases on the other end of that dimension. For the purposes of this analysis, I have assigned both DHHS and DOI releases with scores of -10.0 on the dimension; the EPA releases are scored as +10.0. Given these reference scores, we can, then, assess other environmental agency press releases to determine where they fall between the DOI/DHHS and the EPA. If—as I have argued—PHEPs and NREPs prioritize public health and conservation programs and approaches to a greater degree than environmental protection/enforcement, we should expect that PHEP and NREP texts should cluster around the reference scores of DHHS and DOI, where those agencies prioritize public health and conservation programs and approaches to the greatest degree. We would expect these agencies' scores to differ significantly from that of the EPA documents (i.e., we should be confident that the difference we see between EPA documents and combined environmental agency documents is not simply random).

Data Collection and Calculation of Agency Document Scores

To collect the texts necessary for the analysis, I obtained 2011–2012 press releases from the websites of the DOI, HHS, EPA, and fifteen state-level environmental agencies (see table 6.1 for a list of the states' press releases

Table 6.1. State Agencies' (2011–2012) Press Releases Used in the Analysis by Design Type

Mini-EPA	Natural Resource Combination	Public Health Combination
Arizona Department of Environmental Quality	Delaware Department of Natural Resources and Environmental Control	Hawaii State Department of Health*
Idaho Department of Environmental Quality	Florida Department of Environmental Protection	Kansas Department of Health and Environment
New Mexico Environment Department	New Jersey Department of Environmental Protection	North Dakota Department of Health
Oregon Department of Environmental Quality**	Vermont Department of Environmental Conservation	South Carolina Department of Health and Environment*
Maryland Department of the Environment	Rhode Island Department of Environmental Management	

* 2011, only
** 2012, only

used in the analysis). The fifteen states include four combined public health agencies, five combined natural resource conservation agencies, and six mini-EPA agencies. Agencies chosen for the analysis were chosen randomly from a list of agencies, for which press releases were available and scrapable from agency websites. Once collected, the press releases were combined together into a single document. On average, each state's combined text was approximately 212 pages. These plain text documents do not include any photos or special characters that were included in the original press releases. Additionally, stop words (i.e., words like "and," "to," "or," etc.) were removed from the documents to make the analysis more precise.

After the calculation of relative word frequencies, I specified my reference texts. For the first analysis (table 6.2), I score the document of compiled DOI press releases as -10.0 and the EPA document as +10.0. In the second analysis (table 6.3), I score the document of compiled DHHS press releases as -10.0 and the EPA document as +10.0. Using these scored reference texts, I calculated "wordscores" using Laver et al.'s program. Based on the wordscores, text scores for the each of the virgin texts were calculated, providing raw scores, transformed scores, and confidence intervals. When referencing the enforcement culture scores, I will focus on the transformed scores, as opposed to the raw scores, because the transformed scores are easier to interpret. The virgin text scores are on a different scale than the reference text scores—a scale that is much smaller in comparison. The transformed scores place the virgin texts on the same dispersion metric as the reference texts. Specifically, transforming the scores "preserves the mean and relative positions of the virgin scores but sets their variance equal to that of the reference texts," (316) simply making the scores easier to interpret and compare to one another.

RESULTS OF TEXT ANALYSIS

The state environmental agencies' text scores are presented in tables 6.2 and 6.3. Table 6.2 displays the results of the analysis when DOI is placed at -10.0 on our dimension, while table 6.3 displays the results of the analysis when DHHS is placed at -10.0 on our dimension. The states have been organized by their agency design type to make interpretation of the results clearer. Looking at table 6.2, each of our NREP environmental agencies cluster around one another, receiving negative scores that differ significantly from the nearest non-NREP environmental agency (North Dakota, with a barely positive score of 0.36). NREP environmental agencies, such as Delaware and New Jersey, for example, are indistinguishable from one another, as the confidence intervals overlap. Delaware and the DOI (at −10.0) are also indistinguishable from one another. Florida's NREP scores at −27.03, which places it far beyond the DOI, even. It is important to note that this is certainly possible, given that the DOI's score of −10.0 does not reflect the lowest possible value on our dimension. This simply means that the language used by DOI's press releases is even more prevalent in Florida's releases. Importantly, the NREP agency text scores are clustered around and closer to the DOI's text scores than the text scores of the EPA. This suggests that those program areas, actions, and approaches highlighted by the DOI are more likely to appear in the public language issued by NREPs, as opposed to those highlighted by the EPA. Looking to the other environmental agencies in the analysis, our PHEPs—except for Hawaii—also score relatively low on our dimension (between 0.36 and 2.61), suggesting that the language they are using involves some of what the EPA and DOI use but does not conclusively cluster around one or the other.

Hawaii, scoring fairly close to the EPA at 8.82, is the outlier. The score here means that on a scale where we are able to place the DOI and the EPA, Hawaii's prioritization of programs and approaches more closely mirrors those of the EPA than the DOI. This is not all that surprising, given that Hawaii is a liberal state, that may be less likely to embrace the "conservation for economy" mindset adopted by DOI and many NREPs. Furthermore, New Mexico—a state with a mini-EPA—scores lower than Hawaii, placing

Table 6.2. Text Scores of Environmental Agencies with DOI and EPA Reference Texts

Agency Design Type	State Text	Number of Unique Words Scored	Transformed Score	Transformed Standard Error	95% Confidence Interval		Percentage of Text Scored
Natural Resource Combination (DOI=−10.0)	Florida	5,898	−27.03	0.41	−27.84	−26.22	91.6%
	New Jersey	7,617	−11.97	0.29	−12.55	−11.40	92.9%
	Delaware	7.317	−11.37	0.28	−11.93	−10.80	89.6%
	Vermont	3,707	−8.58	0.62	−9.81	−7.35	92.3%
	Rhode Island	5,772	−6.07	0.31	−6.69	−5.44	90.5%
Public Health Combination	North Dakota	4,069	0.36	0.44	−0.51	1.23	88.8%
	South Carolina	2,981	1.89	0.67	0.55	3.24	88.5%
	Kansas	5,831	2.61	0.32	1.97	3.26	90.5%
	Hawaii	3,061	8.82	0.76	7.32	10.34	90.8%
Mini-EPA (EPA=10.0)	New Mexico	4,331	0.25	0.51	−0.78	1.27	91.7%
	Arizona	3,767	11.64	0.49	10.64	12.63	91.8%
	Idaho	3,423	16.37	0.47	15.44	17.30	95.2%
	Maryland	3,893	17.98	0.59	16.81	19.15	93.9%
	Oregon	4,332	26.54	0.44	25.66	27.42	90.6%

Table 6.3. Text Scores of Environmental Agencies with DHHS and EPA Reference Texts

Agency Design Type	State Text	Number of Unique Words Scored	Transformed Score	Transformed Standard Error	95% Confidence Interval		Percentage of Text Scored
Public Health Combination (DHHS=−10.0)	North Dakota	4,402	−27.92	0.28	−28.49	−27.35	91.8%
	South Carolina	3,114	−16.89	0.45	−17.79	−15.99	89.7%
	Hawaii	3,229	−16.51	0.52	−17.54	−15.48	92.6%
	Kansas	6,270	−11.54	0.23	−12.00	−11.07	92.5%
Natural Resource Combination	Florida	5,931	2.45	0.26	1.93	2.98	91.1%
	Rhode Island	5,847	3.87	0.21	3.44	4.29	90.8%
	Delaware	7,477	5.28	0.19	4.91	5.65	89.6%
	Vermont	3,730	5.56	0.42	4.72	6.40	92.5%
	New Jersey	7,730	5.86	0.19	5.48	6.25	92.7%
Mini-EPA (EPA=10.0)	New Mexico	4,387	5.56	0.34	4.89	6.24	91.8%
	Arizona	3,814	9.94	0.33	9.28	10.60	91.9%
	Maryland	3,960	14.96	0.39	14.18	15.73	94.1%
	Idaho	3,446	18.00	0.31	17.38	18.62	94.8%
	Oregon	4,371	18.97	0.29	18.38	19.56	91.2%

it closer to the DOI. This, too, is not all that remarkable, given that in New Mexico the federal government (via the DOI and Bureau of Land Management) manages around 34.7 percent of New Mexico land. More unexpected is Idaho's relatively high score of 16.37, placing it beyond the EPA's score of 10.0 and further from the DOI, even though the federal government manages around 61.6 percent of the state's land. However, Idaho's situation is a bit more complex. The state did not manage some of its own environmental enforcement programs (e.g., its clean water program) until more recently. Thus, the environmental agency is likely to be more influenced by the EPA than we would expect in a western, natural resource-focused and more conservative state. In general, though, our expectations hold in the text analysis, showing that NREP agencies' language is similar and differs significantly from other agency types, such as PHEPs and mini-EPAs. Furthermore, the NREP language falls closer to that used by the DOI than the EPA, indicating that the programs and approaches prioritized by DOI are more likely to be prioritized by NREPs than the programs and approaches prioritized by the EPA.

In table 6.3, we adjust the dimension by providing scores for the EPA (+10.0) and for DHHS (−10.0). As was the case with NREPs, PHEPs in table 6.3 also cluster around one another, with states like Hawaii and South Carolina, for example, being indistinguishable from each other. That is a somewhat startling result, given the drastic difference in South Carolina's and Hawaii's politics, economies, environmental context, and demographics. And, yet, these two agencies appear to focus on the same kind of programs and approaches, specifically programs and approaches that more closely mirror those emphasized by DHHS. This also holds true for the other two PHEPs in the analysis, North Dakota and Kansas, with scores of −27.92 and −11.54, respectively. Given the new reference texts, NREPs now fall in the middle, not showing definitive preference for either EPA or DHHS programs/ approaches. They appear to look more like the EPA than DHHS but not to the extent of states such as Arizona, Idaho, or Oregon.

This seems reasonable given that NREPs are conservation agencies that would deal only tangentially—if at all—with many of the health issues considered by public health agencies. Another interesting finding, when considering both analyses, is that the environmental agencies' language that is positioned "furthest" from the EPA's language remains relatively far from the EPA, even when new reference texts adjust our dimension. Florida's environmental agency, for example, is the furthest NREP from the EPA, when the DOI's language is used as a reference text *and* when DHHS's language is used as a reference text. This is the case for North Dakota, as well; it is the PHEP that falls furthest from the EPA when both DHHS and DOI are considered. This indicates that the priorities of these agencies are not well-aligned with the

EPA, regardless of the kind of alternative to which they are being compared. These results suggest that there are likely fundamental differences between the programs and approaches preferred by NREPs, PHEPs, and the EPA.

This could certainly cause problems. According to a 2011 EPA Office of Inspector General report on the EPA's oversight of state enforcement, the EPA is unable to "administer a consistent national enforcement program," which means that "state enforcement programs frequently do not meet national goals and states do not always take necessary enforcement actions" (1). The EPA Inspector General's office noted that state enforcement programs were "underperforming" because the EPA is unable to "consistently hold states accountable." Reports like these help underline the results of the analysis I have performed here, which show that states differ significantly in how they think about and prioritize certain environmental protection programming. And, in many cases, states think quite differently about environmental protection than the EPA. Given that the partial preemption process relies on some common understanding of environmental protection between the EPA and the states, this discrepancy in priorities likely creates the kind of conflict between states and the EPA that interviewees in previous chapters described. Citizens in every state are affected when the EPA and states are unable to come to shared conclusions about environmental protection.

Although the analysis shows clear differences between agencies, when we look closely at the results, there are also some puzzling findings, as well. Findings, such as Idaho's text score that places it closest to the EPA, suggest that other factors are determining what agency language looks like in public communications. It is surely not just agency design that determines what kind of programs receive attention and resources and how environmental issues are addressed. Critical, then, is determining if it is the environmental agency design—at all—that has led to these differences, since it is possible that other factors, such as political control or the economy, have made these states similar in their priorities and preferences. Given the extensive number of variables considered in the environmental policy literature, the descriptive text analysis must be considered alongside a number of other factors that we know likely shape the values/priorities of environmental agencies in the states.

HOW DO ENVIRONMENTAL AGENCY DESIGNS COMPARE TO OTHER MOTIVATORS OF ENVIRONMENTAL AGENCY VALUES/PRIORITIES?

While the results in tables 6.2 and 6.3 appear to support the proposition that agency design affects the programs and approaches prioritized by environmental agencies, the visual clustering of these agency text scores is a crude measure of correlation. More simply, it is clear from the outliers discussed previously, and other unclear distinctions, that there are other variables that must be considered in order to state with more certainty that agency design has the effect we expect. We can further investigate this assertion by determining what the effect of agency design is on the programs and approaches highlighted in agency texts (i.e., the measure of agency priorities developed in the previous analysis) when we consider all of the other factors that are likely to motivate environmental agency priorities. Given the theory of environmental agency design, we should expect that the NREP and PHEP combinations should lead to a shift away from EPA programs and approaches and toward those typically seen in conservation and public health agencies, respectively. Thus, I expect that the NREP agency design will result in a decrease in the agency text score, away from EPA agency texts and closer to DOI agency texts, and the PHEP agency design will result in a decrease in the agency text score, away from EPA agency texts and closer to DHHS agency texts. In the first set of analyses, I look at NREP agencies compared to mini-EPAs and PHEPs, and in the second set of analyses I look at PHEP agencies compared to mini-EPAs and NREPs. I also include a number of relevant control variables to account for other factors that shape environmental policy implementation as described in previous chapters.

Briefly, I would like to describe some of the relationships between the control variables I have included and environmental policy implementation. Probably the most widely accepted (both academically and in the minds of citizens) motivator of environmental policy in the states is political control. In particular, we assume that states controlled by Republicans will prefer and require laxer regulatory stringency (Atlas 2007; Hedge and Johnson 2002). In addition to this measure of environmental policy preference, Konisky and Woods (2012b) argue that environmental spending is another powerful indicator. This score may help pick up on a more nuanced scale of environmental ideology that may not be adequately captured by a simple Republican vs. non-Republican score—a score that does not consider how the Republican party's ideology and environmental preferences may differ across the states. In the text analysis, I pointed out that there did appear to be some difference between liberal and conservative states that shaped the scores, with more liberal states still clustered around their common agency design type but falling closer to the EPA than the other combined states. Thus, I expect that those states with unified Republican control should score lower on prioritizing EPA goals and approaches, while those states spending more on environmental protection should score higher.

Additionally, I have suggested in previous chapters that the implementation of laws that limit state environmental agency authority may hinder agencies' ability to execute strict enforcement. Thus, that should be an important control here as well. And, finally, I pointed out that for states like Idaho, the EPA's authority over the implementation of major environmental laws in some states may strengthen enforcement. Therefore, I include these variables in my analyses and expect that they have the suggested effects.[3]

Evaluating the Results

In determining whether the NREP agency design motivates the prioritization of conservation programs and approaches over EPA programs and approaches, we developed a score in the text analyses earlier in the chapter. The clustering of NREP agency text scores we observed around the DOI texts is further reflected and supported by the relationship between the NREP agency design and the agency text scores displayed in table 6.4.

When we evaluate the relationship between the NREP agency design and the text scores produced previously, we can see that the NREP design leads to a decrease in agency text scores away from the EPA and closer to DOI. This indicates that the NREP agency design is associated with a shift away from

Table 6.4. OLS Analysis of NREP Agency Design Effect on Environmental Agency Text Scores, with Controls (2010–2014)

Natural Resource Combination	−10.43**
	(3.30)
Unified Republican Control	2.01
	(2.69)
Environmental Spending	1.21
	(0.76)
Laws Limiting Agency Authority	−14.91**
	(3.88)
Lack of Regulatory Primacy	11.56**
	(4.29)
Number of Polluting Facilities	−0.02**
	(0.01)
State Population	0.07
	(0.31)
State Square Area	62.19**
	(14.36)
Air Quality	−35.31**
	(12.24)
Bureaucratic Capacity	−3.99**
	(0.96)
Change in State GDP	0.37
	(0.40)

n= 70
R^2= 0.77
**<.01

the priorities and preferred approaches of the EPA and toward those that are embraced by an agency dedicated to a natural resource conservation mandate (among other conservation and land management mandates).

We consider this relationship between agency priorities/values and agency design while also considering a number of important state-level factors that might influence the preferences of environmental agencies. However, even when state context is explicitly considered, the NREP agency design continues to push text scores towards the DOI and away from the EPA, suggesting that the NREP agency design is related to a more pronounced focus on conservation programs and approaches. Somewhat surprisingly, neither political control variable appears to have an effect on the agency text scores that we can state with confidence; however, other political factors do appear to matter, including laws that limit an agency's authority, which results in a shift toward DOI programs and approaches and EPA control over particular state water or air programs, which results in a shift towards EPA programs and approaches. This finding helps to explain some of the more puzzling results in the content analysis, where states like New Mexico and Idaho may not have aligned with our expectations. Industry pressure and a stronger bureaucracy also result in a shift toward DOI programs and approaches—those approaches that generally involve less stringent enforcement actions. Certainly a greater industry presence would explain that shift. Additionally, given the research that has found that state-level bureaucrats feel local economic pressure to adjust enforcement behavior (Konisky 2008; Koontz 2002) and the general preference expressed among the bureaucrats in my interviews to work a bit more closely with industry, it is not surprising that when agencies are more capable and powerful, they might exert a preference for more collaborative and cooperative programs and approaches. Air quality presents a more challenging finding with improvements in air quality leading to a shift away from EPA approaches; however, it is certainly possible that improvements in air quality would lead to a lesser need for strict enforcement programs and approaches, overall. In the chapters that follow, I include lags for air quality that help to clarify this complex relationship.

The regression analysis in table 6.4 supports my expectation that the NREP design should move environmental agencies toward those programs and approaches preferred by a conservation agency like the DOI and away from those preferred by the EPA. Table 6.5 displays similar results for the PHEP design; however, there are interesting caveats. Again, the analysis provides additional support for the assumption that the clustering of text scores observed during the text analysis is indicative of a relationship between the PHEP agency design and the prioritization of some programs and approaches over others. In particular, the PHEP agency design moves the agency text scores away from EPA and toward DHHS, indicating that PHEP agencies are more likely to embrace the programs and approaches of agency like DHHS, as opposed to the EPA.

Table 6.5.   OLS Analysis of PHEP Agency Design Effect on
Environmental Agency Text Scores, with Controls (2010–2014)

Public Health Combination	–36.84**
	(5.50)
Unified Republican Control	6.59
	(3.78)
Environmental Spending	1.63**
	(0.78)
Laws Limiting Agency Authority	–6.27
	(3.29)
Lack of Regulatory Primacy	–4.30
	(5.37)
Number of Polluting Facilities	–0.01
	(0.01)
State Population	–0.76
	(0.48)
State Square Area	51.03**
	(12.52)
Air Quality	47.73**
	(13.91)
Bureaucratic Capacity	1.51
	(1.01)
Change in State GDP	–0.70
	(0.41)

n= 70
R^2= 0.76
**<.05

Furthermore, as with the previous NREP analyses, this finding holds, even when other state-level factors are included. When we consider state political, economic, and environmental factors, agency design still appears to shift the priorities of environmental agencies considered in our sample. In fact, the PHEP agency design results in a nearly 37-point decrease in the text score, away from the text score for the EPA at +10.0. According to these analyses, for both the NREP and PHEP design, agency design moves environmental agencies away from including and describing the programs and actions included in the EPA's texts. That being said, this is where the similarity between the two sets of analyses ends. For example, while ideological variables did not appear to be a significant factor when considering the relationship between the NREP design and agency text scores, in the PHEP analyses, environmental spending does appear to play a role in determining the programs and actions highlighted in agency texts. And, the direction of the relationship appears to align with the expectations I expressed previously—that more spending should translate in a closer adherence with the EPA. However, we cannot say with certainty that other political variables, such as laws limiting environmental agency authority or EPA primacy, matter when we are considering the relationship between the PHEP design and an agency's expressed priorities. More important are the size of the state and air quality, where larger states with better air quality are likely to be more aligned with the EPA. While at first glance these differences between the models appear counterintuitive, they actually help to reinforce an important feature of the theory of environmental agency design. As I expressed in the case studies in chapters 3 and 4, we should not expect that the PHEP and NREP designs have identical effects or affect environmental policy implementation in exactly the same way. While we may see laxer enforcement out of each design type, the causal mechanism or *why* and *how* agency design is shifting behavior is not exactly the same. Therefore, where particular constraints—such as primacy or laws limiting agency authority—may matter when we are talking about NREPs, they may matter less or in a different way when we evaluate PHEPs. In the chapters that follow, these differences become even more important, as we consider a number of different kinds of enforcement activities and which of those activities may be most likely to be affected by these unique agency design choices.

That being said, it is possible that the usually powerful political control variables are not significant in our models because political control and agency design are choices that go hand-and-hand. As I noted in chapter 1, researchers analyzing the selection of combined public health and environmental protection agencies noted that political forces helped motivate the decision to adopt or reject a combined public health and environmental protection agency (Sinclair and Whitford 2013). Therefore, it is certainly possible that Republican states are simply more likely to adopt agency designs other than mini-EPAs. In particular, this might be the case for natural resource conservation agencies, where the political control variables in table 6.4 do not even approach statistical significance (in the PHEP analysis, unified Republican control is marginally significant at p<.10). However, a simple correlation test does not support this assertion. Looking at all fifty environmental state agencies between 2010–2017 and comparing Republican control to NREP adoption, we cannot be confident that unified Republican control and the choice to adopt an NREP are correlated (p is $> .10$). Thus, while political control and a state's environmental ideology are important motivators of environmental policy behavior, it does not appear as if agency design is simply another way of measuring environmental ideology. These variables are distinct motivators.

WHAT DOES A MEASURE OF ENVIRONMENTAL AGENCY VALUES REVEAL ABOUT AGENCY DESIGN?

Although abstract concepts, such as an agency's values, institutional preferences, or priorities are difficult to measure, their importance in shaping bureaucratic outputs should not be understated. These internal bureaucratic characteristics play a fundamental role in how bureaucrats perceive their goals, tasks, and approaches to policy problems. In the theory of environmental agency design, understanding how design affects the way agencies prioritize certain programs, tasks, and approaches over others is a fundamental piece of the puzzle in determining how differences in design may eventually also lead to differences in enforcement actions.

While the anecdotes and case studies of chapters 3–5 suggest that combined environmental agency designs may lead to environmental agencies that prioritize conservation or public health programs and approaches over environmental protection, I was unable to compare across environmental agencies directly. Thus, here, we have accomplished two main things: (1) the construction of a score that helps us to evaluate environmental agencies' priorities, comparing them to one another, and (2) an analysis of the relationship between agency design and the expression of those priorities. First, we developed a more concrete way to measure environmental agencies' priorities. To do this, we used the language of environmental agencies. Language is highly reflective of institutional culture and sentiment and, with quantitative content analysis, can help to capture a number of different variables into an interpretable measure. Using Laver et al.'s Wordscores method, I compare the language of environmental agencies to federal agencies that have clearly established priorities and values. With these scores, I was able to establish that combined natural resource agencies speak about their environmental programs in a way that mirrors the DOI, and combined public health agencies speak about their environmental programs in a way that mirrors the DHHS.[4] These agencies differ significantly in their language from mini-EPAs, which more closely mirror the EPA.

Second, upon further dissection of these preliminary findings, I consider the role that other important determinants of environmental policy implementation (e.g., politics, economy, legal constraints) may play in shaping agency priorities. Including these variables alongside agency design in a multivariate analysis, I was able to test our expectation that agency design should affect the way agencies express their priorities and values. Indeed, the findings support that agency design motivates the way agencies speak about their programs and actions. This is the case, even when considering a number of other state-level factors with powerful effects. As was the case in chapters 3–5, agency design appears to play an important and unique role in how agencies view environmental protection.

However, the findings presented here do not establish that agency design directly affects enforcement *actions*. As mentioned early in the chapter, the measure I construct is a measure of values—those feelings, beliefs, and perspectives that likely *spur* actions—not a measure of actions, themselves. It is possible, for example, that although employees of combined environmental agencies state that non-environmental enforcement programs receive more support and resources and text data supports that assertion, the prioritization of public health and conservation does not translate into fewer or less stringent enforcement actions. As one environmental agency employee shared with me in 2011, "just because we lack the support, doesn't mean I don't do my job." For a difference in agency values and priorities to matter, those differences must translate into differences in action.

NOTES

1. Davis et al. (2012) find that corporations, in particular, use press releases to help stockholders make trading choices. By informing stockholders about the direction of the company and future, expected outputs, stockholders can make better decisions about whether or not to increase or lower investments. For public bureaucracies, investment can be translated to political support.

2. I also use this methodology to compare PHEP texts to non-PHEP texts, an analysis similar to the one I perform here (Hopper 2019). Findings are also similar.

3. See the appendix for more detailed descriptions of each variable and the data set I constructed to test the hypotheses presented in chapters 6–8.

4. The measure I constructed here, is, admittedly, not perfect. With abstract concepts, such as agency values, it is difficult to establish a measure that perfectly approximates the concept. However, the text scores are generally affected by each variable in the model the way we would expect. This consistency suggests that the measure likely captures something about agency values and is a potentially valid indicator. It is my opinion that with some refinement, measures, such as the one constructed here, can greatly aid in our ability to understand complex institutions, such as bureaucratic agencies.

Chapter Seven

Do Agency Values
Translate into Actions?

Evaluating the Effects of Agency Design on Enforcement Behavior

In the theory of environmental agency I assume that differences in environmental protection priorities lead to differences in environmental enforcement behavior. In earlier chapters, I describe conversations with environmental agency employees, in which they express that they feel that environmental enforcement is not a top priority and—on occasion—is even discouraged when it involves significant punitive measures aimed at local industries. However, combined environmental agency employees describe political pressures, as well, along with particular characteristics of the state for which they work, that make certain enforcement measures more difficult than others (e.g., states with large square areas to cover by few employees). The question, then, is how consequential is agency design and are agency values when considered alongside other motivators of environmental enforcement behavior? What happens when we compare across fifty states with various political and bureaucratic structures, ideological preferences, and economic conditions? Do values and priorities translate into discernible actions?

To begin to answer this question, we, first, have to evaluate some of the complexities inherent to measuring and predicting environmental enforcement behavior. As Konisky and Woods (2012b) note in their thorough survey of environmental policy indicators, it is more complex than we might imagine. Measures, such as the number of inspections an agency performs, a state's dependence on natural resources, or state citizens' involvement in environmental interest groups can all be indicators of some kind of environmental context, but these measures are not always correlated strongly or in the direction we might assume (Konisky and Woods 2012b). For example, states in the American West, such as Oregon and Washington, are economically dependent on natural resources, such as timber. Based on this assumption, alone, we might assume that the states cater to industry and exhibit laxer enforcement behavior. However, Oregon and Washington are also liberal states that have regularly endorsed environmental policy actions that push beyond federal standards. Conversely, states like Wyoming and Montana are considered conservative states, with a strong history of "rugged individualism" that helps ensure the maintenance of freedoms for industry that push against environmental protection goals. However, these two states have a relatively high number of citizen members of the Sierra Club, one of the largest environmental organizations in the United States (Konisky and Woods 2012a). This suggests that these states' governments should be under pressure from citizens to pursue more progressive environmental policies. These contradictions exist because the landscape of factors that drives environmental policy is full of mediating and moderating factors that produce a unique environmental policy outcome in each state. These differences among the states are one of the things that makes studying environmental policy in the American states beneficial, as it allows us to make more generalizations than studying federal environmental policy actions, alone. However, this complexity has also led us to focus on some motivations of environmental policy in the states, at the expense of others.

Of course, one area in which the environmental policy literature has produced less nuanced descriptions and findings is in the distinction between the set of factors that drives the content and passage of environmental statutes and the policies of elected officials and the set of factors that drives the creation of environmental rules, regulations, and the enforcement of those rules and regulations. Certainly, these factors overlap; however, as Meier and O'Toole (2006) and Maynard-Moody and Herbert (1989) argue, the shape, function, and process of bureaucracies are important determinants of policy implementation. And this certainly makes sense, given that bureaucrats are the individuals with most direct consequence for day-to-day implementation that are not necessarily open to the direct influence of political principals. However, political science's soft continuation of the politics-administration dichotomy has often led us to think of bureaucracies as extensions of their political principals, rather than tremendously political institutions in their own right, where structures, values, and other organizational factors affect policy implementation choices. More simply, jumping directly from governors and state legislatures to environmental enforcement behavior ignores the power that bureaucracies wield over regulatory policy implementation, and this may lead us to a less coherent understanding of the complex web of factors that affect environmental policy in the states.

This acknowledgment is important in understanding and assessing the theory of environmental agency design because the theory relies on the assumption that the nature of environmental agencies has a direct effect on environmental enforcement behavior, even given the political, environmental, and economic context of a state. We expect that the organizational context within which environmental regulators operate shapes their behavior. Also, the recognition that not all environmental policy indicators are mutual indicates that we must be cautious in how we choose to measure "enforcement behavior." We have to closely consider what kind of behaviors would be most and least likely to be affected by bureaucratic characteristics.

ENVIRONMENTAL ENFORCEMENT BEHAVIOR

In the NREP and PHEP case studies, I describe a number of allegations made against MODNR, NYDEC, KDHE, and CDPHE in regard to their leniency with local, polluting industries. In particular, many of these allegations deal with accommodations made to regulated entities. Industries are allegedly not held accountable for violations and are under-penalized when they are held accountable. Although these combined agencies all publicly embrace innovative approaches to helping industry to cooperate with environmental standards, they are all criticized for cutting corners in terms of enforcement. Although combined agencies argue that enforcement is not always necessary to ensuring compliance, their agency performance is still being assessed by environmental/industry interest groups, citizens, and the EPA, using enforcement actions as an indicator of dedication to environmental protection efforts. This is something we observe clearly in the case studies in chapters 3–5.

Agencies are judged by their enforcement actions because enforcement actions have played a significant role in environmental protection efforts since the focus of those efforts became centered on regulating industry in the 1970s. The command and control approach, advocated by the EPA, is a regulatory approach in which regulatory standards are set in accordance with national law, and the EPA enforces those standards through regular monitoring and penalization for the violation of standards. An important foundation of the command and control approach is that while environmental standards can be "established with reference to incremental control costs," "environmental laws rarely give EPA this discretion" (EPA 2016a). More generally, this means that the cost of regulations posed to industry is not to be considered the priority during the rule-making process; the effects on human health should be considered as the top priority. This is in direct contradiction to many of the regulatory philosophies adopted by combined state agencies, in which "feasibility" is regularly assessed and prioritized. This distinction is important to note because it extends far beyond the rule-making process.

A pivotal part of the command and control approach is the set of enforcement actions used by regulators to address industries that violate regulatory standards. The EPA labels the enforcement of environmental laws as a "central part of EPA's Strategic Plan to protect human health and the environment" (EPA 2016b). Enforcement actions are assessed by the EPA or state environmental agencies upon the assessment of a violation(s) of environmental regulations. For the EPA, "cost to industry" is not the most important component of rulemaking, and it is also not the most important component of enforcement. The EPA calls for "tough civil and criminal enforcement for violations that threaten communities and the environment" (EPA 2016c). Cooperation, flexibility, or partnership with industry is not listed among the EPA's (2016c) "Enforcement Goals." The EPA does mention using "innovative enforcement strategies to improve compliance"; however, placing emphasis on keeping industry costs low is in direct opposition to the regulatory philosophy put forth by the EPA.

In sum, there are two important components of the EPA's perspective on regulation that likely influence their use of enforcement actions: (1) enforcement actions are an important part of the regulatory process championed by the EPA—command and control—and (2) the EPA's regulatory process does not always take the "cost to industry" under consideration as seriously as other evaluative criteria, such as health. However, because the EPA and state environmental agencies often separately assess violations and implement enforcement actions, enforcement actions are one of the areas in environmental policy where states have the most discretion (Atlas 2007). More simply, states may penalize violating facilities much less than the EPA suggests is appropriate, and this is acceptable under the partial preemption system because as long as states are meeting baseline environmental standards, *how* they meet those standards is within their discretion. In fact, even when states do not meet environmental standards, it is very unlikely that the EPA will intervene by taking away state primacy, as studies have found that federal oversight is lax and varies greatly across the states. "The EPA has never suspended a state's authorization for failure to meet federal requirements, even in cases where states clearly are not enforcing the federal regulations as envisioned" (Konisky 2008, 323; see also EPA/OIG 1998, 2005; GAO 2000, 2006). Therefore, because there are no real federal requirements for state-level enforcement actions—only strong suggestions—the level and severity of those actions varies by the state.

WHAT CAUSES DIFFERENCES
IN ENFORCEMENT BEHAVIOR?

The existing environmental policy literature has explored a number of determinants of state-level environmental policy. States may base their environmental policies on need, matching environmental policies to pollution problems (Sapat 2004; Daley and Garand 2005; Lowry 1992), or states may make environmental policies that reflect a pro- or anti-environment citizenry (Hays et al. 1996; Daley and Garand 2005; Woods 2008) or environmental/ manufacturing interests within the states (Ringquist 1993; Hays et al. 1996; Bacot and Dawes 1997; Davis and Davis 1999; Daley and Garand 2005). And, finally, the political control of state government may determine what environmental policies look like, with Democratic control leading to stricter environmental policies and Republican control leading to more lenient environmental policies (Wood 1992; Woods 2008; Atlas 2007; see also Hedge and Johnson 2002). These are all accepted explanations for the variation in environmental policies that exists across the American states.

However, as I mention above, the theory of environmental agency design suggests that other factors, such as the organizational characteristics of state environmental agencies, likely have an even more direct effect on the day-to-day enforcement behavior that defines environmental policy implementation in the states. While we have long accepted and found evidence that a state's political context helps shape its regulatory behavior, we have also regularly assumed that bureaucracies are simply an extension of that political control. As I relay previously, this is problematic because it fails to acknowledge the powerful organizational values, structures, rules, traditions, and norms that drive the actions of regulators (Maynard-Moody and Herbert 1989; Meier and O'Toole 2006). Furthermore, while bureaucratic actions are likely affected by the pressures applied by elected officials, things like agency values and the general direction of agency policy implementation are much less likely to be penetrable to political pressure (Ringquist 1995). The preferences of elected officials must filter through a highly structured, rule-bound, and path-dependent process before those preferences are translated into policy implementation actions. This is seemingly like a game of telephone, where the organization molds and shapes policy directives depending upon their own preferences, capabilities, and understanding of what constitutes "best practices." By the time a statute reaches the ground, it may look quite different in action than it was in intention.

Additionally, the path-dependent nature of bureaucratic agencies ensures that these relatively impermeable organizations become even more so, as they continue to endure ever-changing political regimes, environmental problems, and technology. Even if changes in the organization of these agencies could help open them up to political officials, these political reorganizations are costly and contentious. Thus, as the world changes around environmental agencies, established ideas, preferences, and values continue to drive the agency to pursue implementation along a "path of least resistance," and the path of least resistance is often the path that does not deviate far from the well-established trail. Once settled upon this path, agencies then adjust their direction based on the impediments introduced by economic and political pressures. As state-level economic pressures tend to more acutely affect those regulators working in state-level environmental agencies (Konisky 2008), we should expect that any agency path leading toward more collaborative and cooperative approaches with industry during the regulatory process would be further reinforced by these pressures. For NREPs and PHEPs, agency preferences for more flexible regulatory approaches should, thus, perpetuate in the face of *and* be potentially compounded by political and economic pressures.

As outlined above, enforcement tools are an important part of environmental protection programming as expected by the EPA; however, states can and do differ in their enforcement behaviors because of the discretion provided to states on how to meet federal environmental standards. If, as previous chapters suggest, the combined agency design serves to elevate the programs and approaches of natural resource conservation and public health at the expense of the programs and preferences of environmental protection, states with combined agency designs should provide more flexible and cooperative approaches to industry during the regulatory process. This should occur because public health and natural resource conservation agencies tend to endorse more flexible approaches to regulation, and regulation may be made more difficult by the traditional structuring of program/service delivery (e.g., the county-by-county approach of public health). Costs to industry as a result of enforcement actions should be taken into consideration for states where environmental agencies emphasize flexibility and cooperation with industry. This means the assignment of lower penalties and enforcement actions taken during the regulatory process. Thus, I expect that a combined agency design will lead to lower annual penalty amounts and fewer enforcement actions taken per number of violations discovered. This should occur, even while controlling for other important factors that have been evaluated in the previous literature, including economic, political, and environmental context, and other factors that are not always included in these models, including the existence of laws limiting agency authority, bureaucratic capacity, and previous years' penalizations. These were factors that my case studies suggested were important determinants of enforcement behavior, in addition to agency design.

DATA AND ANALYSES

The first challenge in assessing whether design affects enforcement behavior is in deciding how to measure enforcement behavior. I choose to operationalize the use of enforcement tools by using two variables that measure areas in which states have quite a bit of discretion—whether they determine a violation merits an enforcement action(s) and the severity with which states choose to penalize violators. I measure these outputs by the total annual dollar amount of penalties assigned to polluters by state environmental agencies and the annual number of enforcement actions taken, given the number of violations discovered and documented by state environmental agencies.

Total penalty amounts reflect variation in the stringency of enforcement measures because states use a "variety of written penalty policies to determine what penalty [the state] should seek in settling a case, and also what its 'bottom line' for the penalty amount will be in settlement discussions with a violator" (EPA 2016d) The penalty outcome is directly dependent on how states choose to measure unfair economic benefits and environmental harm, and each state measures these considerations in their own way. Additionally, all states use monetary penalties as an enforcement tool, as the EPA labels penalties an important way to "level the playing field" among businesses, preventing industries from profiting from environmentally dangerous practices.

The number of enforcement actions assigned by regulators introduces us to another part of the regulatory process. Specifically, identifying a violation is the first step in an investigation process that can eventually lead to formal enforcement actions, such as the administration of civil monetary penalties. As the Government Accountability Office (GAO) (1991) notes, the EPA relies on "strong and effective . . . state enforcement to deter violations and to help ensure the timely correction of identified violations." However, because states have historically had limited resources, regulators are "unable to take enforcement actions against all violators," which leads environmental agencies to pursue those enforcement actions they think most consequential (GAO 1991, 6). That being said, the ability of state agencies to "determine when non-compliance is 'significant' enough to warrant an enforcement action," also allows environmental agencies a considerable amount of discretion in determining if enforcement actions should be taken at all (ibid.). Thus, this is another part of the enforcement process in which state environmental agencies are able to provide flexibility to regulated entities if they see fit *or* if they are unable to act with stringency, due to a lack of resources and support. It is crucial to note that the use of two dependent variables that measure enforcement activity during different parts of the enforcement process is necessary, given that the PHEP and NREP agency designs affect enforcement in different ways (as illustrated by previous case studies and analyses).

A Base Model of Environmental Enforcement Behavior

To show how the theory of environmental agency design adjusts our understanding of how powerfully certain factors affect enforcement behavior, we

Table 7.1. Common OLS Analysis of Environmental Enforcement Behavior (Annual Penalty Amounts, 2010–2017)

Unified Republican Control	−391072.40*
	(228767.30)
Number of Polluting Facilities	180.19
	(141.64)
Population	77517.20**
	(16895.59)
Square Area (logged)	1802.68
	(102965.30)
Air Quality	1855107.00*
	(1002400.00)
League of Conservation Voters Average Score	−3822.03
	(4389.85)
Bureaucratic Capacity	137881.50**
	(60680.53)
Change in GSP	−78273.47*
	(44350.52)
Constant	−1589412.00*
	(904725.70)

Adjusted R2= 0.11
n= 400
p-value: * <.10
** <.05

can first look to a base model of environmental enforcement behavior (table 7.1) similar to those models that have been used previously in state environmental politics research. Of course, this model is not reflective of each study's specification; however, it includes many of the variables that scholars of environmental politics consider most consequential in predicting environmental enforcement behavior (see previous description of determinants of environmental enforcement behavior in this chapter and chapter 1).

In this model, which uses total annual penalty amounts as its measure of enforcement behavior, we can see that political variables, environmental conditions, and bureaucratic capacity are all important factors in determining the severity of enforcements. In the case of political control, unified Republican control should result in around a $391,000 decrease in total penalties for the year, as compared to states in which Democrats control some part of state government. Additionally, a growing economy, as measured by changes in state gross product, also leads to a decrease in penalty amounts, potentially helping to further boost successful industries that support the states' economies. Factors that increase penalty amounts and enforcement stringency include population (more people = more pollution), air quality, and bureaucratic capacity. Here, because I have not included a lag for air quality yet, it is difficult to know whether higher penalty amounts or better air quality occurred first. In regard to bureaucratic capacity, the more capable environmental agencies are, the higher penalties we tend to see. This is in accordance with existing research (Carpenter 2001; Nicholson-Crotty and Miller 2012) that indicates that stronger and more respected bureaucracies tend to benefit from some political insulation due to their power and capacity. Furthermore, in my interviews with bureaucrats, agencies, and departments that stated they had support from their state legislatures and governors tended to claim they had more authority over their day-to-day enforcement actions.

Results when Including Agency Design Variables

In table 7.2, I include two variations of an agency design measure—a simple dummy variable for the combined agency design and variables that break that variable down further by design type (NREP, PHEP, and mini-EPA). In Model 1, I use a variable that indicates whether or not a state has a combined environmental agency. Additionally, I include a few other measures that elaborate on the base model, including environmental spending, an air quality lag, a state's economic dependence on natural resources (likely to be especially important for states with the NREP design), laws limiting agency actions, enforcement primacy, and the previous year's total penalty amounts (measured by a lagged dependent variable). These variables are also included in Model 2. In Model 1, the combined agency design type results in a nearly $644,000 decrease in total penalty amounts in comparison to the mini-EPA design type. This decrease in penalty amounts is substantive, given the average annual per capita penalty amounts that we see by state in the dataset (see appendix 2). This finding supports the theory of environmental agency design, in which we expect that the combination of environmental enforcement with another policy area should result in a minimization of environmental enforcement programs than those approaches preferred by the EPA. And, we find that the combined agency design type matters, even when we consider the power of political control, where governors and legislators may be pushing for a particular approach to dealing with industry. Other factors that we can be confident matter include the number of polluting facilities, which pushes penalty amounts higher as more facilities tend to mean a higher number of violations, and air quality. Once again, bureaucratic capacity—as measured by states' score on a range of management "grades," which include management of resources, staff, information, infrastructure, and so forth—is likely to increase penalty amounts. This seems reasonable, given that agencies' programs and preferences are likely much more influential when the bureaucracy is a powerful actor, working alongside the state legislature and

Table 7.2. Multi-Level Random Intercept Analysis of Combined Agency Design and Environmental Enforcement Behavior (Annual Penalty Amounts, 2010–2017)

	Model 1	Model 2
Combined Agency Design	–643763.20**	—
	(205322.10	
NREP Agency Design	—	–816005.00**
		(225488.60)
PHEP Agency Design	—	–195398.10
		(321373.70)
Unified Republican Control	–47089.33	–70772.51
	(248627.90)	(247821.70)
Number of Polluting Facilities	414.16**	403.83**
	(152.16)	(151.57)
Environmental Spending	347.62	289.311
	(275.16)	(275.77)
Population	22190.72	33260.67
(in millions)	(25891.43)	(26488.90)
Square Area (logged)	–66881.55	-99676.61
	(125026.80)	(125762.40)
Air Quality	2163803.00*	2096593.00*
	(1270962.00)	(1265615)
Air Quality (t-1)	259390.10	267108.70
	(1305506)	(1299460.00)
Bureaucratic Capacity	164840.80**	182123.90**
	(65843.34)	(66131.72)
Change in GSP	–51118.69	–69110.92
	(56152.91)	(56771.78)
Natural Resource Dependence	–7636.10	–10166.24
	(127063.10)	(126481.70)
Laws Limiting Agency Actions	125006.50	150024.50
	(199621.90)	(199177.70)
No Enforcement Primacy	–167615.50	–96307.02
	(307557.90)	(308662.80)
Sierra Club Membership	122675.00	120517.40
	(130450.80)	(129851.40)
Annual Penalty Amount (t-1)	0.05	0.05
	(0.05)	(0.05)
Constant	–2723116.00**	–2833718.00
	(1164511.00)	(1160725.00)

n= 350
p-value: * <.10
** <.05

governor. That said, if it is true that environmental agencies' values differ, then how capacity affects enforcement behavior should also differ (a proposition I deal with directly in chapter 8).

The results presented in Model 2 (table 7.2), contain a number of similarities to the results in Model 1. The primary difference is the effect of agency design, which has now been broken down by agency design type, with mini-EPAs being the excluded comparison category. In comparison to those mini-EPAs, NREPs appear to produce lower penalty amounts, decreasing annual total penalties in NREP states by around $816,000. In this model, while the coefficient for PHEPs is also negative (as we would expect), we cannot be confident that the negative effect PHEPs appear to have on penalty amounts is not simply a random finding. Thus, we can say with certainty that only the NREP design appears to have the effect on penalty amounts that we might expect.[1] Why NREPs and not PHEPs? First, our lack of confidence in the PHEP finding does not mean that the design does not affect penalty amounts. Case studies and interviews with a number of PHEP employees suggest that it likely does. Instead, it is possible that the effect of the PHEP and NREP designs differ in output. For example, our case studies suggest that the simple addition of a public health mandate to environmental protection mandates can create structural problems that lead to diminished enforcement capacity and sensitivity to economic effects. This could mean that the inspection and violation discovery process is more difficult. Also, many PHEP employees and the PHEP case studies suggested that many of these agencies have mechanisms in place to encourage more informal enforcement measures that may never make it to the penalty stage. Regardless, our understanding of the complexity of enforcement decisions indicates that the PHEP and NREP design types likely affect enforcement behavior in various ways, so this distinction between the two is not unexpected. As with Model 1, the number of polluting facilities, air quality, and bureaucratic capacity also continue to affect enforcement behavior, increasing penalty amounts to varying degrees. However, agency design and bureaucratic capacity appear to have the most substantive effects on enforcement behavior in which we can be confident.

Looking now to table 7.3, I incorporate an alternative measure of enforcement behavior—the number of enforcement actions taken (either formal or informal), per the number of violations discovered and documented by environmental agencies. As I note above, this is one of the first steps in the regulatory process, and it is a measure that demonstrates how willing or capable agencies are of pursuing enforcement actions for all violations of environmental rules and statutes. As with the models presented in table 7.3, Models 3 and 4 also incorporate measures of political control, economic context, environmental context, and bureaucratic capacity. I also continue to include controls for the previous year's air quality and the previous year's number of enforcement actions.

Table 7.3. Random Effects Analysis of Combined Agency Design by Type and Environmental Enforcement Behavior (Enforcement Actions Taken per Number of Violations Assigned, 2010–2017)

	Model 3	Model 4
Combined Agency Design	−1.15**	—
	(0.59)	
NREP Agency Design	—	−1.07
		(0.71)
PHEP Agency Design	—	−1.35**
		(0.71)
Unified Republican Control	−1.83*	−1.82*
	(1.02)	(1.03)
Number of Polluting Facilities	−0.01	−0.01
	(0.01)	(0.01)
Environmental Spending	−0.01	−0.01
	(0.01)	(0.01)
Population (in millions)	0.03	0.03
	(0.08)	(0.07)
Square Area (logged)	−0.12	−0.11
	(0.36)	(0.35)
Air Quality	10.29**	10.30**
	(5.31)	(5.31)
Air Quality (t-1)	−15.04**	−15.03**
	(4.24)	(4.24)
Bureaucratic Capacity	0.04	0.03
	(0.36)	(0.37)
Change in GSP	−0.02	−0.01
	(0.16)	(0.17)
Natural Resource Dependence	−0.14	−0.13
	(0.37)	(0.37)
Laws Limiting Agency Actions	0.22	0.21
	(0.56)	(0.56)
No Enforcement Primacy	−0.23	−0.26
	(1.03)	(1.07)
Sierra Club Membership	−0.11	−0.11
	(0.58)	(0.58)
Enforcement Actions Taken per Number of Violations Assigned (t-1)	0.08	0.08
	(0.07)	(0.07)
Constant	7.44	7.48
	(4.91)	(4.97)

n= 350
p-value: * <.10
** <.06

In Model 3, we can see that the combined agency design type leads to a decrease in the number of enforcement actions taken per violation by more than a single enforcement action. Once again, given that most states' annual enforcement action rate ranges between zero and five actions per violation in the data (see appendix 2), this appears to be a substantive decrease. It also suggests that the combined agency design has the effect that we expect; it results in fewer enforcement actions, even when violations are discovered by inspectors. This provides evidentiary support for our earlier findings that combined agency regulators may be more reluctant to take enforcement actions, choosing instead to work with the industry informally to help them back into compliance. Additionally, it can indicate (as the GAO suggests) that these agencies struggle logistically to pursue the appropriate number of enforcement actions for all violations discovered. In addition to the combined agency design type, it appears that political control also affects the way in which agencies apply enforcement actions. Given the results in Model 3, we should expect that a state environmental agency under unified Republican control would take almost two fewer enforcement actions per violation. This means that the agencies may be dissuaded from or not provided enough support or resources to pursue enforcement, even when violations are uncovered. The regulatory process does not even make it to a determination of how severe a penalization should be for a violating entity. The third factor that appears to drive enforcement behavior is air quality, with both the current and previous year's air quality having a powerful effect on the number of enforcement actions that are taken per violation.

In opposition to the findings in the previous models (table 7.2), however, a bureaucracy's management capacity does not appear to have an effect when controlling for other economic and political factors. It is possible that this finding is the result of political officials having more control over how capable bureaucracies are of pursuing enforcement actions in the first place. Legislatures and governors that are opposed to more stringent regulation may seek to provide fewer resources to environmental agencies, which would make it difficult to distinguish between a state agency with conservative political leadership and a state agency that lacks bureaucratic capacity; they could be one in the same. Unfortunately, because the management "grades" used to create the bureaucratic capacity score apply to a state's entire bureaucracy, we are unable to determine with certainty whether diminished environmental agency capacity and unified Republican control are strongly correlated. That said, our case study of the TCEQ provides a clear example of the number of control mechanisms that exist and can be used by state legislatures and governors to diminish the power and capacity of environmental agencies. Therefore, it is likely that environmental agency capacity and political control are related in some or all states.

In Model 4, I break down the agency design type to help us understand any differences between PHEPs and NREPs. While the penalty model shows that we can be confident only in the NREP agency design type's negative effect on penalty amounts (i.e., severity of enforcement actions), the results in Model 4 suggest the opposite. Here, we can be more confident that the PHEP agency design's negative effect on the number of enforcement actions taken per violation than we can in the NREP agency design's negative effect. As was the case with the previous models, NREP continues to have a negative effect; however, it falls just outside of marginal statistical significance, with a p-value of 0.13. Again, given our findings in the case studies that show NREP employees and agencies seem reluctant to pursue enforcement actions for each and every violation discovered, I would expect with more data that we would gain enough confidence in the NREP design's negative effect. That being said, it is also possible that we are more confident in PHEP's effect on enforcement actions because—as I have noted previously—the structuring of public health programs and policy implementation may make it more difficult for agencies to perform thorough inspections, discover violations, and pursue enforcement actions for each violation. Resources and expertise may be spread thin across public health departments performing these regulatory actions. For NREPs, conservation programs are less likely to be structured on a county-by-county basis, so we may not see as powerful an effect at this point in the process. Instead, as Model 2 (table 7.3) suggests, we may be more likely to see NREPs express a willingness to bargain with industry on penalty amounts—a practice that emerged as an important part of the NREP regulatory process in both the Missouri and New York case studies. As with Model 3, we must also consider the role of strong Republican leadership and environmental conditions, which also appear to affect states' decision making about how to pursue enforcement actions for violations.

The broad assumptions of the theory of environmental agency design continue to be supported, even as we observe a different stage in the enforcement process. The enforcement and compliance data that states have submitted to the EPA provide additional evidence that combined agency design types lead to laxer enforcement behavior. This provides further support for the findings in case studies and interviews where NREPs and PHEPs appear to favor more collaborative and cooperative approaches within industry that keep economic interests in mind and seek to improve compliance through working closely with industry to help find a common middle ground. Additionally, the findings here complement the complexity that we uncovered in those case studies and interviews; PHEPs and NREPs approach enforcement differently, but those differences may be due to varying factors. Public health and natural resource conservation agencies do not generally house officials with the same educational backgrounds, employ the same structures for implementation, or rely on the same sources for revenue. They pursue mandates that overlap via environmental protection but are otherwise isolated from one another. In fact, as the 1960s debates over a potential EPA's structure reveal, one of the reasons for creating an independent environmental agency was to address the very real differences in goals and approaches between the public health and natural resource agencies that were controlling environmental protection. Therefore, we should not expect that these agency design types affect enforcement behavior in the same way. Rather, the theory of environmental agency design insists simply that design type and the mandates that are combined with environmental protection are powerful motivators of the day-to-day enforcement decisions of regulators. State-level enforcement behavior supports this proposition, even when considering the numerous other factors that environmental policy scholarship finds relevant. The inclusion of bureaucratic characteristics, including agency design choices, is imperative to understanding environmental enforcement behavior in the states.

However, these findings continue to leave us with a number of questions. The appearance and disappearance of political effects and bureaucratic capacity in these models is not inexplicable given the theory of environmental agency design, but it brings to light an important possibility: in some cases, some institutions may matter more than others. The variation across the states in regard to environmental agency design type is not the only variation that has the potential to effect environmental protection in the states. How powerful are these Republican legislatures that seek to promote more economically sensitive regulation? How powerful are the mini-EPAs that may fight against them? The balance of power in the states may be of similar structure to our federal government (executive vs. legislative vs. judicial), but how power is distributed among the branches varies greatly depending upon the state. If this is the case, in some states, environmental agency design may be less consequential, simply because a powerful governor or legislature has the ability to directly drive the behavior of bureaucrats. Thus, in refining the theory of environmental agency design, the acknowledgment of differences in bureaucratic and political power is critical.

NOTE

1. This could also be due to the small number of states in our dataset that have the PHEP agency design, making estimation more difficult.

Chapter Eight

Organizational Capacity, Agency Design, and Environmental Enforcement

"They don't really know what they're doing," an environmental regulator shared with me when I asked them if they felt supported by the state legislature. "We can usually just wait them out if they try to cause drama." This was not the first or last instance, in which environmental regulators would share that their political principals lacked the capacity to control them. Often environmental agency employees would indicate that political appointees heading their agencies were either a temporary nuisance or a temporary source of support for important programs. However, on a number of occasions, I would also hear the opposite: that agencies and their employees were terrified of pushing back too much. They did not want to bring the kind of attention to their agencies that could result in less funding and cuts to staff. Employees were keenly aware of a balance of power that could determine the fate of their programs and jobs. Some were more frightened by what they saw than others.

As I discuss in the early chapters of this book, the politics of each individual state play an important role in what environmental enforcement looks like. For example, in California, a highly professionalized Democratic legislature is likely to reinforce a focus on stringent regulation, while the powerful executive appointments of Texas Republicans may hinder the ability of TCEQ to enforce environmental statutes without directly incorporating the interests of industry. The existing research on the political control of bureaucracies suggests that a tangled set of factors determines how much autonomy bureaucracies like environmental agencies yield, including how effectively law may have been constructed to insulate bureaucracies from political control (Lewis 2004), whether political officials are concerned about an agency's actions, and how easily political principals may be disadvantaged by information asymmetries and conflicts between executives, courts, and legislatures (Hammond and Knott 1996). However, even when bureaucracies hold relatively little autonomy, no one political institution drives agencies' behavior entirely (Hammond and Knott 1996). A web of competing actors produces unique policy outputs in each state. Thus, understanding how the structure and power of political institutions in the states vary in their effects has important implications for the implementation of environmental policy. Furthermore, if we are to fully understand how consequential agency design may be for environmental enforcement behavior, we must also consider how the strength of other state institutions may mediate an environmental agency's ability to assert its preferences.

There is quite a bit of state institutional variation to consider. For example, in some states, governors may have veto or budgetary powers that governors in other states lack. Governors' power to appoint may be limited or expanded. Legislatures can be made up of citizens who receive little compensation for their work and meet infrequently, or they can be made up of skilled politicians with full working staffs and a salary that allows them to make legislating a career (Squire 1992; Squire 2007; Squire 2012). And, as the management "grades" that make up our measure of bureaucratic capacity suggest, bureaucracies differ too. How many staff can they train and pay? How satisfied are their employees, and how frequently do those employees enter and exit the organization? Where do they receive their funding? In the sections that follow, I elaborate on some of the differences in capacity across state legislatures and bureaucracies relevant to environmental policy implementation.

LEGISLATIVE CAPACITY

American state legislatures changed drastically over the course of the twentieth century, as massive expansions of the administrative state and increasingly complex public problems encouraged the development of more efficient and effective representative bodies. This growth in efficiency and effectiveness (particularly in the *substance* of policy making) is labeled as professionalization, measured by organizational attributes such as more pay, longer session lengths, more staff assistance, and better facilities (Squire 1992; Squire 2007). These changes incentivize members to stay longer in the legislature, allowing them to develop more policy expertise and providing members with the informational resources necessary to craft more complex and innovative policy solutions (Squire 2012). In fact, the number of bills passed in the legislature and the number of bills enacted per legislative day increases with professionalization (Squire 1998). This, along with "increased membership stability and diversity, greater policymaking capacity, and increased responsiveness," comprise an impressive number of positive implications accompanying increasing professionalization in state legislatures (Woods and Baranowski 2006, 585; Rosenthal 1996; Squire 1992; Thompson 1986).

One of the other motivators of legislative professionalization was an effort to strengthen state legislatures vis-à-vis massively powerful executives and organized interests (Woods and Baranowski 2006). The hope was to create a more educated, diverse, and capable legislature that could act independently from governors and state agencies (along with special interests). And, in accordance with reformers' intentions, "professional legislatures" appear to be "better able to counter gubernatorial influence in the budget process to better resist the governor's policy agenda and to more effectively constrain the bureaucracy" (Squire 2012, 311; Huber, Shipan, and Pfahler 2001; Kousser and Phillips 2009; 2010). Specifically, the increase in staff helps to provide legislators with the informational resources necessary to narrow the informational advantage enjoyed by career bureaucrats or specialists (Squire 2007; Rosenthal 1996, 171–72).

However, others have argued that the existence of these additional resources does not necessarily ensure their proper use. Other evidence suggests that indicators of more professionalized legislatures are negatively associated with the ability to influence executive agencies (Elling 1984). Specifically, "more capable legislatures often exert less influence" (Elling 1984, 167). Similarly—and related to environmental agencies, specifically—others find that legislative professionalism has no significant effect on the influence reported by state clean air agency directors, as of 1998 (Potoski and Woods 2001). The complexities in these findings are likely due to the multi-dimensional nature of professionalism; greater institutional resources (such as an increase in staff assistance) can and does enhance legislative influence, but the move toward careerism that also accompanies professionalization largely diminishes influence (Woods and Baranowski 2006). "Career oriented legislators tend to be more focused on electoral concerns," and oversight simply does not get the attention of the electorate. In their own analyses, Woods and Baranowski (2006) find support for this theory, determining that while two of three components of professionalization—legislative staff and operating budget—significantly increase the influence of the legislature, other components actually decrease legislative influence. Therefore, while legislative professionalism is a well-documented measurement for legislative capacity, in the case of a legislature's capacity for oversight, the number of staff available to legislators appears to have the greatest impact on oversight capabilities.

Another important variable that likely affects whether or not legislators are capable of oversight is legislative turnover. Short stints within the state legislature can have a number of negative consequences for effective and efficient policy making, along with performing oversight. The implementation of term limits, which increases turnover by default, has brought about significant changes in the power dynamic between the executive and legislative branches. The relationship is notably asymmetric, as governors have gained more authority at the expense of term limited legislatures (Carey, Niemi, and Powell 2000; Carey et al. 2006; Moncrief and Thompson 2001; Powell 2007). For instance, after the implementation of term limits in Michigan, 19 percent of the state's House saw the governor as the most influential member of the body; whereas, that number sat at 6 percent prior to term limits (Sarbaugh-Thompson et al. 2004). This makes sense because although the majority of governors are also subjected to term-limits themselves, the information asymmetry that already exists between legislative principals and bureaucratic agents is further widened as novice legislators attempt to oversee and negotiate with the executive branch.

However, it is not only governors that will see an increase in authority because of informational advantages (Carey, Niemi, and Powell 2000; Carey, Niemi, Powell, et al. 2006). Bureaucrats—overall—are thought to see an increase in power, further suggesting that as legislative institutions are "debilitated by loss of expertise" and "term limits damage the policymaking and policy oversight capacities of legislatures relative to other institutional actors," the executive becomes powerful in its ability to direct and determine the content of policy in the states (93). As it relates to environmental enforcement, the informational gap between specialized bureaucrats within a highly technical policy area and legislators is already wide. Thus, impeding upon the ability of legislators to develop expertise in this area likely has an even more significant effect within environmental policy than it may have in other policy arenas.

Both professionalization, particularly the number of staff available to legislators, and term limits are institutional characteristics of legislatures that define the capacity of the body to influence its bureaucratic agents. As I note above, environmental protection is a highly specialized and technical area of policy for which the learning curve is steep. Those legislators without access to staff and without the time necessary to become knowledgeable about potential policy solutions and challenges are likely at the disposal of experts within the bureaucracy. Additionally, bureaucrats may have even more of an advantage over the day-to-day enforcement actions because legislative influence is least prevalent in decisions made about organizational structure, daily operations, and personnel (Elling 1984). However, when these types of decisions are made poorly by bureaucracies, they may not be able to wield as much influence.

BUREAUCRATIC CAPACITY

Like legislatures, states differ significantly in their bureaucracies' expertise, information processing, innovativeness, and efficiency (Barrilleaux, Feiock, and Crew 1992). Advances in technology, degree requirements, representation, salary levels, and more, all vary across state bureaucracies, posing important consequences for policy implementation. Investing in strengthening bureaucratic capacity—as defined by some of these characteristics—is worthwhile for bureaucratic agencies because it allows an agency "to determine up front the kinds of expertise, personnel, and data that can be brought to bear on specific policies" (Ting 2011, 246). Legislatures may feel less capable of challenging bureaucracies through the oversight process, when legislators are keenly aware of a "technological disadvantage." Thus, capable bureaucracies likely wield more influence over the implementation of policies.

In addition to technological advantages, higher levels of bureaucratic capacity are also likely to encourage legislatures to invest more trust in bureaucratic processes. Bureaucracies with reputations as powerful and competent actors receive more support and more trust from political principals (Carpenter 2001). This enables bureaucratic agencies to more effectively lobby political actors and push for support for bureaucratic preferences (Nicholson-Crotty and Miller 2012). Thus, the overall management capacity of the states—including financial management, human resources management, managing for results, capacity management, and information technology—mediates the level of influence bureaucracies have in the policy-making process (controlling for a number of other factors) (Nicholson-Crotty and Miller 2012).

For environmental policy, especially, technological advantages and trust may be even more consequential. As agencies that engage with a highly technical policy area, environmental agencies already hold an informational advantage over most political principals; however, the regulatory nature of environmental agency policy implementation and the skepticism aimed at those regulatory actions may lessen any power gained from that informational advantage. Thus, the perceived effectiveness of state bureaucracies is likely a crucial factor in whether environmental agencies are able to pursue their regulatory preferences.

HOW MIGHT DIFFERENCES IN INSTITUTIONAL CAPACITY AFFECT ENVIRONMENTAL POLICY IMPLEMENTATION?

In the states, a patchwork of checks and balances determines how strongly environmental bureaucracies can assert their preferences. Due to partial preemption, which dictates that the EPA may delegate enforcement power to the states, state institutions are powerful environmental policy actors. And, the implications of state actors' environmental policy decisions do not stop at state borders. Thus, analyzing those differences is vital in determining what factors most powerfully shape enforcement behavior and environmental quality in the United States. What, then, are some of the possible environmental policy implications of this variation in state institutional capacity?

As I have discussed in previous chapters, existing research and conventional wisdom states that more conservative political principals, such as state legislatures in Republican-controlled states, push for less stringent environmental enforcement, while more liberal political principals push for more stringent environmental enforcement. In regard to legislative professionalization, professionalization has the ability to either strengthen or weaken the oversight powers of legislatures. Thus, if Republicans were to control a more professionalized legislature, we may expect two potential outcomes: this control in a stronger, professionalized legislature could lead to more stringent environmental enforcement measures, as those Republican legislators seek to pursue the kind of activities that are more electorally relevant and rewarding than oversight, *or* Republican legislators in a professionalized legislature may be empowered, through staff, pay, and time in the legislature to perform oversight more effectively. This would allow them to more effectively push for less stringent enforcement measures. In regard to term limits, much of the existing literature suggests that this forced turnover greatly weakens legislatures in respect to the executive, including bureaucratic agencies. Thus, term limits should lead to Republican legislators' weakened ability or propensity to develop expertise and perform oversight of environmental agencies. The capacity for management within a state's agencies, or bureaucratic effectiveness, also has the potential to affect environmental agencies' strength or the trust that legislators place within the bureaucracy. Specifically, as bureaucratic effectiveness increases, we should see the agency become able to more aggressively pursue its preferred style of regulation as described by the theory of environmental agency design.

To evaluate these possibilities, we can look to some of the same data I have used in previous chapters, specifically 2010–2014 environmental enforcement data,[1] including our indicator of environmental agency design, and a compiled data set of measures of legislative partisan makeup, legislative staff, legislative term limits, agency employee numbers, bureaucratic effectiveness, and other relevant control variables that I have shown in previous chapters to be theoretically important predictors of the rates and severity of environmental enforcement actions.

Once again, we can measure enforcement stringency by using the sum of total penalty amounts that are levied against violating industries. As I note in chapter 7, the EPA claims that states use a "variety of written penalty policies to determine what penalty [the state agency] should seek in settling a case, and also what its 'bottom line' for the penalty amount will be in settlement discussions with a violator." Considerations when determining penalties include the economic benefit that violators may have gained from their noncompliance and the environmental harm caused by noncompliance with environmental statutes. Thus, the outcome of penalty negotiations is directly dependent upon how states choose to measure unfair economic benefits and environmental harm, and each state measures these considerations in their own way. However, in addition to this, penalization decisions also capture agencies' willingness and ability to bargain with industry during the penalty determination process—bargaining power that may be more or less prevalent in states that have chosen a particular agency design type. In sum, states are given a considerable amount of discretion in determining the severity of environmental enforcements, and it is this variation among the states in how they choose to issue monetary punishments that makes this measure a reasonable indicator of a more or less stringent preference for strong environmental enforcement. Lastly, monetary penalization is generally the most public and adversarial show of force by an environmental agency (i.e., the type of enforcement action most likely to garner the attention of political principals and special interests). Thus, it is likely that influence over this enforcement action, as opposed to something like required daily inspections, is most likely to be pursued by legislators or representatives of industry.[2]

To measure the partisan makeup of state legislatures, I use a dummy variable, indicating which political party controls the upper and lower chambers—counting those chambers that are divided in their partisan control as zeroes. A one for this variable indicates that Republicans have complete control over the legislature in the state. This is fairly common, with around 44 percent of legislatures controlled completely by Republicans over my time span. As I describe above, Republican control—particularly of the legislature—has been shown by previous studies (such as Atlas 2007) to decrease penalty amounts issued by environmental agencies. Furthermore, analyses in chapter 7 suggest that unified Republican leadership may lead to fewer enforcement actions per the number of violations discovered, which is related to weaker enforcement behavior, overall. This also aligns well with the previous cases, in which Republican-controlled states like Missouri, Kansas, and Texas tended to lean toward more cooperative approaches with industry, even when organizational decisions, economic variables, and environmental variables differed.

Importantly, these measures of enforcement behavior and political control are the same measures we have relied on in previous analyses. The difference here is that we need to adjust those relationships by considering the relative strength and capacity of institutions. Thus, I look to two measures of legislative capacity to help account for these differences across state institutions: the number of legislative staff available to state legislators and whether the state's legislature is subject to term limits. I use legislative staff rather than using the entirety of Squire's legislative professionalization index, due to the multi-dimensional nature of professionalization noted in the existing research on legislative capacity (Woods and Baranowski 2006). Specifically, increases in staff are thought to provide legislatures with increased capacity to perform oversight, while other parts of professionalization, such as higher pay, may act to increase career ambition, which may lead to even less focus on oversight.[3] I also consider whether the state's legislature is subject to term limits. There are some complications in using term limits as a predictive measure because term limits implemented in 2000 or later may not yet have had a significant impact on the legislatures in my time sample (2010–2014). However, a simple indicator for whether or not a state legislature is subject to term limits is commonly used within the term limits literature, even given these limitations.

In addition to considering factors that may determine how effectively state legislatures may control bureaucratic behavior, we also must consider the ability of bureaucracies to push back or avoid interference altogether. This kind of durability in bureaucratic agencies should allow for agency design to have the strongest effect possible. Since we are evaluating how competing regulatory preferences between state institutions are affected by the relative abilities of state institutions, it makes sense for us to look specifically at an environmental agency design that would cause a Republican state legislature the most trouble. According to our case studies and analyses thus far, mini-EPAs are the most likely to take a command and control approach to regulation that prioritizes stricter enforcement actions. This is in line with the agency after which they are designed—the federal EPA.

A major difference between these agencies and the EPA, though, is that even though state-level mini-EPAs share a similar organizational design, they vary drastically in their size, political and social support, and resources. Therefore, they differ, too, in their ability to execute the pollution control mandate and the competence needed to express their regulatory preferences without interference from the legislature or special interests during that process. Measuring bureaucratic capacity is a difficult ask, given the number of factors that go into determining how susceptible bureaucracies may be to outside interference; however, the Maxwell School's GPP overall management capacity grades for the states helps to combine a number of factors that shape bureaucracies' durability (Nicholson-Crotty and Miller 2012). The measure provides letter grades for the states, in regard to their information technology, financial management, human resources management, managing for results, and capacity management capabilities. The measure helps to capture a bureaucracy's assumed competency and should indicate how much trust is placed in the agency to carry out its mandate in line with its own preferences.[4]

Results of Institutional Capacity Analyses

In table 8.1, I use enforcement data to evaluate the effects of institutional capacity on our understanding of the implications of environmental agency design.[5] Both models in table 8.1 include the effect exerted by the mini-EPA agency design and Republican control of the legislature. The models differ only in the combination of mediating factors included for the state legislature. Model 1 shows the effects of the mini-EPA agency design mediated by management capacity and Republican control of the legislature, as mediated by the number of staff available to legislators. Model 2 shows the effects of mini-EPA agency design mediated by management capacity and Republican control of the legislature, as mediated by term limits.

In Model 1, both relationships with mediating factors appear to have a statistically significant effect on penalty amounts issued by state agencies, as do the penalty amounts issued in the previous year, the number of polluting facilities monitored by the EPA, and the state's air quality. This would indicate that institutional capacity likely adjusts the translation of the regulatory preferences expressed by Republican-controlled legislatures and mini-EPAs into enforcement actions. As was also the case in chapter 7, a state's air quality and the number of polluting facilities are important motivators of enforcement severity, as well. However, when institutional capacity is integrated into our analyses, other factors emerge as being consequential, as well. For example, the state's dependence on natural resources is marginally significant (at $p<.10$), and Republican control—once mediating effects are considered for both bureaucratic influence and legislative influence—emerges as a significant factor. The annual penalty models in chapter 7 do not indicate a significant political control effect, which I note is surprising given the previous literature's strong findings related to this variable. However, once we consider how *strong* both the legislatures and bureaucracies are in the states, we can be more confident in the effect of political control. Theoretically, this is reasonable because it is difficult to determine that particular enforcement behaviors are due to the regulatory preferences of one institution or another without understanding which institutions have the

Table 8.1. Legislative and Bureaucratic Influence on Environmental Enforcement Actions, with Mediating Factors

	Model 1	Model 2
Penalty Amounts (t-1)	0.195**	0.237**
	(0.056)	(0.067)
Mini-EPA Agency Design	−936.607**	−890.017*
	(435.428)	(475.199)
Bureaucratic Effectiveness	−150.038**	−120.746*
	(50.968)	(63.330)
Mini-EPA Agency Design x Bureaucratic Effectiveness	226.919**	208.067**
	(86.424)	(94.936)
Republican Control of Legislature	−474.817**	−32.300
	(199.415)	(121.935)
Legislative Staff	−0.162	0.396**
	(0.227)	(0.192)
Term Limits	131.266	44.270
		(254.677)
Republican Control x Legislative Staff	0.715**	—
	(0.205)	
Republican Control x Term Limits	—	138.044
		(205.292)
Seats in the Legislature	0.019	0.327
	(0.869)	(0.846)
Sierra Club Membership	−9.913	4.477
	(49.326)	(59.334)
Number of Polluting Facilities	0.306**	0.216
	(0.140)	(0.144)
Population	0.004	0.002
	(0.018)	(0.019)
State Square Area	−0.001	−0.001
	(0.001)	(0.001)
Spending	−0.001	−0.001
	(0.001)	(0.001)
Air Quality	−1787.281**	−1647.756**
	(762.891)	(803.865)
Air Quality (t-1)	219.756	45.823
	(835.592)	(883.285)
Dependence on Natural Resources	137.748*	104.443
	(72.446)	(69.570)
Poor Health Days	−1.431	69.075
	(68.093)	(68.843)
Change in Gross State Product	18.903	23.735
	(18.986)	(18.191)
Laws Limiting Agency Authority	-- 188.848	−176.724
	(149.444)	(152.321)
Lack of Primacy	349.192*	252.738
	(200.011)	(174.050)
Constant	1988.559**	1354.589
	(772.694)	(873.756)

Beta coefficients are listed, with standard errors in parentheses. GLS random effects regressions, with clustered standard errors by state.
n=250
P< 0.5** <.10*

ability to exert their preferences over policy implementation. Specifically, legislatures are not directly making enforcement decisions, so without knowing how much power they may have over bureaucratic outputs, it is difficult to connect legislative preferences to those bureaucratic outputs.

To interpret the interaction terms further, we can look to figures 8.1 and 8.2, which show substantive effects. In figure 8.1, we can see the effect of the mini-EPA agency design at different levels of management capacity, as measured by the Maxwell School's GPP overall management capacity grades for the states. These grades range from one to eight, with one being a failing grade and eight being an excellent grade. In figure 8.1, we see that at low levels of bureaucratic capacity, mini-EPA penalty amounts decrease. This provides support for our expectation that low levels of bureaucratic effectiveness may weaken an agency's ability to pursue its own regulatory preferences, when considering Republican control. Since mini-EPAs tend to preference more stringent penalization (in line with the command and control regulatory style of the EPA), this preference is likely to suffer when the agency is less effective. Looking back to our case study of the Texas Commission on Environmental Quality (TCEQ), this finding helps to illustrate how a state with a mini-EPA could continue to exert relatively weaker enforcement behavior.

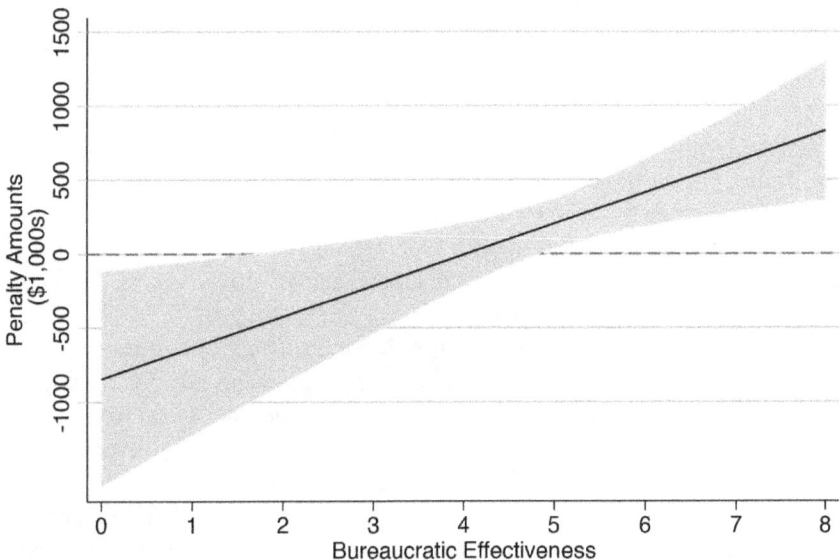

Figure 8.1. The Effect of Mini-EPA Design at Different Levels of Bureaucratic Effectiveness on Penalty Amounts.
Created by author, using Stata and an author-compiled dataset of state-level variables. See appendix 2 for summary statistics and data sources. See appendix 3 for more detailed descriptions of variables.

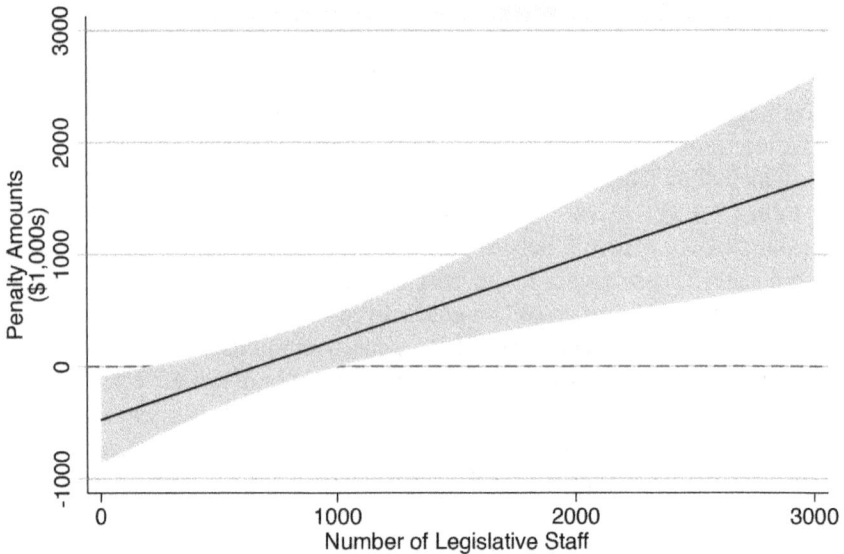

Figure 8.2.   The Effect of Republican Control at Different Levels of Legislative Staff on Penalty Amount.
Created by author, using Stata and an author-compiled dataset of state-level variables. See appendix 2 for summary statistics and data sources. See appendix 3 for more detailed descriptions of variables.

In Texas, where the environmental bureaucracy is constrained by law and has a negative reputation among lawmakers, weakened management capacity likely affects the ability of the agency to exert strong enforcement prefer-ences—preferences that our case study shows *do* exist. Conversely, as is the case for the powerful California Environmental Protection Agency (CalEPA), we should expect to see that a highly effective and reputable environmental agency would be able to exert a preference for strong enforcement measures, based in the command and control ideology of the EPA. This does appear to be the case, as figure 8.1 shows that as agencies' effectiveness increases, penalty amounts increase, as well.

   In figure 8.2, we can see the effect of Republican control at different levels of legislative staff (i.e., different levels of legislative capacity for oversight). Here, low levels of legislative staff (in presumably unprofessionalized legisla-tures) do not have a significant effect on penalty amounts. However, as staff increases to over one thousand (controlling for the number of seats in the leg-islature), we see that penalty amounts actually increase. This does not provide support for the assumption that increased staff support should help Republicans to exert their regulatory preferences more effectively over the environmental enforcement process. Instead, it provides support for the alternative expectation that state legislators working within highly professionalized legislatures, even when isolated to only staff, have weakened motivations to perform oversight due to oversight activities' relatively thin electoral rewards. This allows environmental agencies to act without much oversight during the enforcement process, which would allow mini-EPAs to pursue more stringent environmental enforcement without as many active constraints by the legislature. Of note, here, is that Model 1 suggests that when Republican legislatures are less professionalized—they tend to be less professionalized than legislatures controlled by Democrats, on average—we should expect that they act to decrease penalty amounts (around $415). This finding is in accordance with previous research that evaluates the relationship between political control and environmental enforcement. However, with the consideration of agency design and the mediating effects of institutional capacity, we can see an important distinction—that the effects of political control are shaped by institutional factors. This should not be surprising, as the theory of environmental agency design suggests that institutional factors are an important determinant in institutional behavior.

In the remaining model shown in table 8.1 (Model 2), I exchange legislative staff numbers for an alternative measure of institutional capacity: term limits. Previously, I posited that term limits should weaken Republican legislatures' oversight capabilities, leading to more stringent environmental enforcement. However, the results in Model 2 do not allow us to be confident that term limits are leading to more aggressive enforcement actions in environmental agencies. Instead, factors that have been powerful in previous analyses, such as previous enforcement actions, legislative staff numbers, and air quality appear to exert more convincing effects when term limits are considered. Additionally, this model shows that legislative staff, when not considered as a mediating factor of party control and while controlling for term limits as a mediating factor of party control, appears to have a positive and statistically significant effect on penalty amounts. This is interesting because it suggests that had we simply used legislative staff as a control variable, we might assume that the effect on penalty amounts was uniform across party control, rather than understanding that an increase in professionalism and potential decrease in oversight activities would lead to less desirable actions taken by the environmental agency relative to the party in control. In figure 8.3, for example, we can see the effect of the mediating relationship between legislative staff and party control, considering Democratically controlled legislatures. Here, the opposite effect occurs, where Democrats would see a decrease in penalty amounts with higher staff numbers, a scenario that is more likely to occur because highly professionalized legislatures are more likely to be Democratic. Without considering the interaction, we would not have a clear understanding of how professionalization affects a legislatures' ability to exert its regulatory preferences on environmental agencies. Also, in both models, the mediating relationships

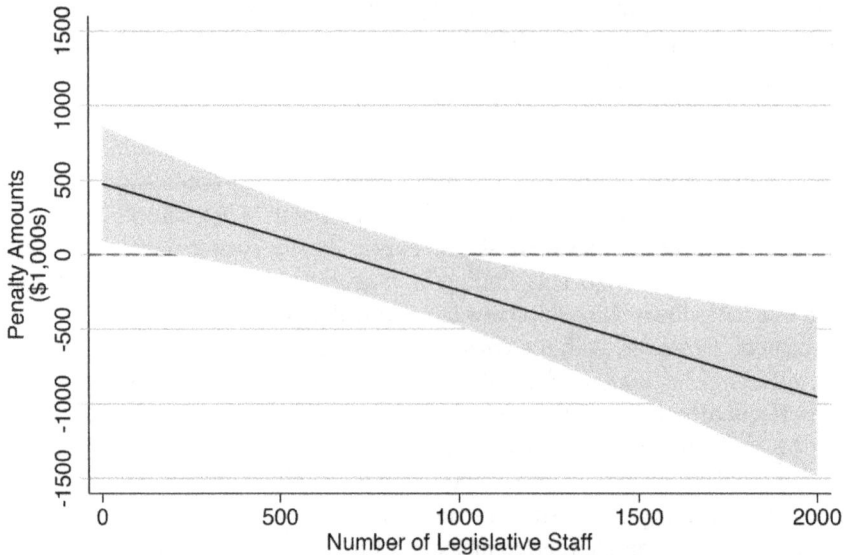

Figure 8.3. The Effect of Democratic Control at Different Levels of Legislative Staff on Penalty Amount.
"Created by author, using Stata and an author-compiled dataset of state-level variables. See appendix 2 for summary statistics and data sources. See appendix 3 for more detailed descriptions of variables.

between agency design and bureaucratic effectiveness matter, *even* as term limits or the strength of the legislature are considered.

THE IMPLICATIONS OF INSTITUTIONAL CAPACITY FOR THE THEORY OF ENVIRONMENTAL AGENCY DESIGN

Analyses of enforcement data suggest that institutional characteristics—particularly those that help us to measure an institution's capacity for pursuing its preferences in opposition to the preferences of another institution—likely mediate the influence that bureaucracies and legislatures have over environmental enforcement in the states. The nature of these findings also illustrates how nuanced those mediating relationships may be, as some factors (such as professionalization) have surprising effects, while others (such as term limits) fail to produce a relationship in which we can be confident. A measure of bureaucratic effectiveness produces effects that we might expect—that a more capable bureaucratic agency is better able to exert its regulatory preferences.

It is of some interest that a variable such as term limits would not render an effect on penalty amounts due to the crippling effects term limits have on legislatures' ability to compete with powerful executive branches. As I mentioned previously, it is possible that term limits have not yet been in place long enough to produce the kind of effects that we expect or that oversight is such a "lowly" activity for legislators that stunted time within the legislature makes no real difference in how oversight is executed. Another possibility is that the correlation between professionalization, political control, and the application of term limits may make term limits' effects difficult to estimate. As stated earlier, many Republican controlled legislatures are less professionalized than Democratic controlled legislatures, such as California, New York, and Illinois. This brings up another consideration, which, if it is rare to have a highly professionalized legislature, then we should expect that—most of the time—unified Republican control has a negative or null effect on penalty amounts. It is only in those relatively few instances where Republican legislatures are professionalized or have large staff numbers, where we should see the inversion take place, where Republicans may actually end up with larger penalty amounts than expected. This may actually be much more problematic for Democrats, who—as figure 3 shows—would experience a substantive decrease in penalty amounts at higher levels of professionalization—a scenario that is more probable. For example, a state like New York, with a professionalized legislature and a combined environmental agency may be even more likely to produce more lenient enforcement than another state with a combined environmental agency and Democratically controlled legislature (e.g., Hawaii).

It should also be noted that there are other institutional characteristics, aside from staff numbers and management scores, that could be used as indicators of capacity. For example, another measure of professionalization is legislator pay. Although this does not serve the same theoretical function as staff—explicitly and directly helping legislators to collect and synthesize information—it is a measure that has an identical counterpart on the bureaucracy side: government worker pay. Given the difficulties in assessing management grades for environmental agencies, specifically, using pay may be an alternative. However, its limitations in measuring the capacity for oversight in legislatures should be weighed against its advantages. Additionally, these analyses also do not consider the power/capacity of state courts or governors, which could also have a significant impact on the implementation of environmental policy in the states. To fully specify models of environmental enforcement in the future, these are all factors that should be included.

Those limitations considered, these analyses suggest that we should be considering the full scope of institutions that affect environmental policy implementation and the relative power of those institutions within the states. There are challenges to incorporating these factors correctly (e.g., ensuring that we accurately measure and represent institutional preferences for environmental enforcement); however, our current understanding of environmental enforcement behavior in the United States is incomplete without considering the push and pull between state institutions.

As I describe at the beginning of this chapter, environmental agency employees and the history and guiding documents of environmental agencies reveal unique organizational norms, personalities, and—as J. Q. Wilson puts it—"the ways things are done around here." In the previous chapters, we have evaluated how agency design may help to shape these organizational personalities and actions. However, how able agencies are to insulate those norms and how easily those norms are translated into actions is a notable variable in determining how agencies choose and are able to regulate. Agency design can be directly connected to enforcement behavior only if organizational personalities are able to persist and influence the day-to-day decisions of regulators. Therefore, organizational capacity matters, and it matters for every institution that seeks to control the direction of policy. Thus, if we are asking who and what exerts power over environmental policy implementation in the states, the answer to that question is that it depends. It depends on the state's economic and environmental conditions, its citizens' attitudes toward the environment and regulation, environmental agencies' and political officials' preferred approach to regulation and enforcement, and how powerful and capable bureaucracies and political officials are in opposition to one another. In every state these relationships produce a unique version of environmental protection. Depending upon where you live, you might live longer because of it.

NOTES

1. Due to the static nature of the bureaucratic capacity variable (i.e., the measure was calculated in 2008 and has not been updated since), I feel most comfortable looking only to years closest to 2008 in my dataset (only about five years out). Thus, I have chosen not to use my full dataset, which spans to 2017 for the analyses in this chapter.

2. Penalty amounts are also used as indicators of enforcement stringency by Atlas (2007), Hopper (2017; 2019), and others.

3. I use the National Conference of State Legislature's estimates of legislative staff for "total staff during the session" for 2009. Additionally, it should be noted that data also exist for 2015, the year after my time span ends, which may be a more accurate staffing point for the years, such as 2013 and 2014. This being said, due to what appears to be significant changes in staffing for some states between 2009 and 2015, I

was reluctant to assume which years these changes occurred in, choosing to utilize more conservative estimates of staff.

4. It is important to note that this measure is not agency-specific, so it is possible that a state's environmental agency may be presumed incompetent by the legislature, while agencies—overall—are trusted. Controlling for Republican preferences should help to address some of this concern; however, the possibility that this measure may incompletely capture environmental agency competence should be considered carefully when interpreting these results. The creation of individual measures of environmental agency competence are outside of the scope of the analyses in this book.

5. To model the effects of these mediating factors on agency/legislative influence over environmental enforcement, I use a generalized least squares (GLS) model with random effects and standard errors clustered on the state. As the nature of the data (cross-sectional time series) and diagnostics suggest that ordinary least square would be inefficient, GLS may be used to account for heteroscedasticity in the errors (Wooldridge 2009, 278). Given the potential for (or really, certainty of) state-specific effects, fixed effects would be the ideal method of analysis. However, a number of the variables in my models do not vary across the time span I consider, including some of my primary independent variables, such as environmental agency design, legislative staff, and term limits. Additionally, my use of a theoretically important lagged dependent variable could also be problematic with fixed effects, due to the potential for Nickell bias. The likelihood that this bias would exist is even higher, given my data's short time span (Nickell 1981; Hsiao 2003). Thus, I use random effects, and I include the lagged dependent variable, which helps to account for some of the remaining autocorrelation.

Chapter Nine

The Implications of a Theory of Environmental Agency Design

The theory of environmental agency design is based on a simple premise: that the combination of environmental enforcement with other policy implementation mandates affects the way agencies view regulation and may alter their ability to actively address violations of environmental statutes. In the preceding chapters, I have laid out evidence to support the following assertions. First, NREPs tend to embrace more collaborative and cooperative approaches with industry, particularly during penalty determination; this is not far removed from the historical tendency of conservation agencies to focus on the state/local economic outputs of resource extraction and a reliance on continued resource extraction for agency funding. Interviews suggest that employees of natural resource agencies are encouraged to work closely with regulated entities and to adjust enforcement measures to help protect the economic well-being of those entities. Voluntary compliance is the ideal.

For PHEPs, there also appears to be a tendency toward more leniency with regulated entities, but it is not for the same reasons that we see as influential for NREPs. As state public health agencies have taken on a more prominent role as health service and insurance provider, they have shifted away from the more regulatory functions latent to environmental health programs (Kotchian 1997). Additionally, PHEPs are likely spread too thin, given the growing number of mandates they are expected to serve, forcing inspectors and regulators to choose carefully among environmental protection efforts. The local nature of public health policy implementation and funding also increases the potential for local economic pressures to penetrate day-to-day regulatory work. As can be the case with all combinations of policy mandates, a less politically popular and more difficult set of agency tasks likely suffers when placed alongside relatively anything else. In this case, environmental enforcement is simply not as politically or socially popular, and, according to PHEP employees they often lack the staff and resources to fully pursue enforcement goals. This was often also the case for NREP employees as well, albeit for sometimes different reasons.

Quantitative analyses of environmental agency documents and enforcement measures across the states revealed that environmental protection programs likely do suffer when combined with public health or natural resource conservation, with non-environmental programs or less regulatory environmental programs receiving more attention, support, and action. This minimization of environmental programs appears to translate into differences in enforcement behavior, as well as with combined environmental agencies punishing fewer violations with enforcement actions or negotiating lower penalty amounts for documented violations. Agency design continues to have this negative effect, even when controlling for the political, economic, and environmental context of a state. Examinations of the relationship between politics and agency design do not suggest that agency design is simply a mirror for the party in power. Agency design choices have an important and independent effect on the way that agency employees behave.

That being said, the power of agency design is limited by the powers invested in and wielded by legislatures and bureaucracies in the states. A mini-EPA in a state where bureaucracies are excellent managers of implementation with plenty of resources and support is more likely to be able to translate their preferences for stricter regulation than those mini-EPAs in states with weaker bureaucracies. For legislatures, the effects of political control are mediated by institutional characteristics, like the number of staff, that determine how aggressively legislators can and want to pursue oversight of bureaucratic agencies. These mediating factors suggest that our understanding of environmental enforcement behavior in the states is dependent upon our full understanding of the balance of power in state governments and the potential for each political actor to penetrate the day-to-day choices of agency personnel. One thing, though, is arguably certain: our models of environmental enforcement must acknowledge the importance of the norms, structures, and values of the environmental agencies that directly execute enforcement actions. Factors like agency design matter.

While the premise of the theory of environmental agency design is simple; its implications are more complex given the number of considerations it introduces to theories of environmental policy implementation, broadly speaking. While I have focused primarily on the most direct actions coming out of environmental agencies as my measure of environmental policy implementation, we still have to ask, why would more collaborative and cooperative approaches with industry matter? How could this change human exposure to pollution, ecological health and diversity, and human and species health?

Are combined environmental agencies good or bad for the environment? The answers to these questions are not simple, nor are they definitive; however, there are a few things that we do know: (1) environmental agencies across the board are moving toward more industry-inclusive approaches; however, those approaches are most successful when strict environmental enforcement continues to play a role; (2) environmental issues are growing more complex and severe (e.g., the expansive climate change crisis), and attention to and support for environmental protection is vital in abating crises; (3) and the American states continue to expand their role in environmental protection, making any differences in environmental protection even more consequential for citizens within a state and for any and all citizens who may indirectly suffer the consequences of another state's actions. According to the World Health Organization, outdoor air pollution causes ischaemic heart disease, strokes, chronic obstructive pulmonary disease, lung cancer, and acute lower respiratory infections in children (WHO 2014). And, research shows that "water, air, and soil pollution causes 40 percent of deaths worldwide" (Lang 2007). Previous research suggests that more protective policies in the American states lead to better environmental conditions (Konisky and Woods 2012a; Ringquist 1993; Ringquist 1995; Woods et al. 2009). In the discussion that follows, I briefly address each of these points.

In 2006, Bennear argued that working with, as opposed to against industry, during the regulatory process could help to increase levels of compliance and to, subsequently, limit harmful pollution or the exploitation of natural resources. As I mention in previous chapters, it is in the best interest of regulators to have positive working relationships with industry, as they are dependent upon industries' transparency and honesty, in order to regulate effectively. Including industry in the regulatory process or choosing to use management-based approaches or market incentives, as opposed to strict command and control, can help improve relationships between regulators and regulated entities, making regulators and the regulations they impose more efficient and productive (Harrington and Morgenstern 2007). Thus, it should not be surprising that environmental agencies, including the EPA, are relying more heavily on non-command and control approaches to regulation. For example, the 1990 amendments to the Clean Air Act that created a cap-and-trade system for regulating sulfur dioxide and oxides of nitrogen replace strict command and control measures with a market-based system that helps cap pollution at safer levels and allow those industries that rely most heavily on the production of those pollutants to make smaller adjustments than those industries that are less reliant. Additionally, as recently as 2018, the EPA has followed the lead of some states by encouraging self-disclosed violation policies, which allow "regulated entities to voluntarily discover, promptly disclose, expeditiously correct, and take steps to prevent recurrence of environmental violations," with some immunity from punishment (EPA 2018). These are just two of a number of examples of programs/policies that the EPA has enacted in a push toward more industry inclusive regulation.

However, with this shift in policy has also come a drop-off in enforcement. According to an Associated Press news article from February 2019, the number of civil investigations carried out by the EPA dropped from 125 in 2016 to 22 in 2018, and penalty assignment declined from about $207 million in 2016 to $86 million in 2018 (Knickmeyer 2019; Konisky and Woods 2018).[1] While there has been a "general trend downward" in enforcement for "many years," numbers in 2018 portrayed a twenty-five-year low. While this could simply be a reflection of a shift toward more collaborative and industry-inclusive approaches to regulation (and not a direct reflection of environmental protection laxity), evidence suggests that strong enforcement may be a necessary precursor and complementary tool for non-command and control approaches. For example, in order for management-based approaches that allow for more cooperation and flexibility with industry to work effectively to curb violations, some level of enforcement must remain (Coglianese and Lazer 2003; Hopper 2017). At the extreme, if there is absolutely no possibility of punishment for non-compliance, it is unlikely that industry would see it in their best interest to take on potentially costly changes in production and disposal methods. Thus, as a steady and reasonable threat of enforcement actions looms, industries are more likely to choose the route of voluntary compliance, which would allow them to have more say during the regulatory process. Continued enforcement is vital in helping to apply necessary pressure.

And, pressure for compliance is necessary because environmental problems continue to plague human and ecological health, and are growing only more difficult to address. The Ash Council that advised Nixon on the creation of the EPA in the late 1960s could likely not anticipate the number, diversity, and complexity of environmental issues that the agency would eventually face—in particular, the threats posed by global warming. According to the National Oceanic and Atmospheric Administration (NOAA) (2019), unprecedented human-caused global warming is posed to cause severe drought followed by devastating flooding, poorer water quality, the spread of disease and famine, the elimination of species, and coastal degradation. Unmitigated climate change will be expensive, and it will be deadly. Unfortunately, we have already begun to experience some of these effects (NOAA 2019). The scale of the problem requires a considerable amount of global and state/local cooperation from citizens and governments who face a variety of economic and political challenges. Furthermore, according to the 2018 Global Climate Change Assessment, we have reached a point, in which the problem requires immediate changes in policy and the enforcement of those policy changes to abate disaster. If a decline in enforcement leads to less effective environmental protection efforts, the shift toward more industry-inclusive approaches could be potentially damaging. Additionally, environmental protection's continued secondary status in combined environmental agencies can lead to lack of funding, resources, and support that renders regulatory programs ineffective at solving the highly complex problems regulators must face.

Of course, the EPA is not a combined environmental agency, so some might argue that the issue continues to have solid representation. However, states and localities continue to grow in their ownership of environmental policy implementation and enforcement. For example, upon the Trump administration's withdrawal from the Paris Climate Agreement, states and cities have become active in undertaking their own efforts to address climate change. Former California governor Jerry Brown reached out to the nations that are part of the Paris Climate agreement to help arrange how California could participate in the agreement, even though the United States had formally withdrawn support. But, the shift toward greater power for states/localities does not always, or even generally, lead to more progressive environmental protection or enforcement. For example, the Trump administration's recent endeavors to delegate greater environmental enforcement power to states and localities has resulted in less stringent enforcement measures and a persistent deregulation effort, as evidenced by the administration's efforts to eliminate the Clean Power Plan and to adjust the Waters of the United States (WOTUS) rule (Konisky and Woods 2018).

Although it is difficult in the short-term to determine the effects of these devolution and deregulation efforts, one thing is for certain: states have more environmental enforcement power. And, if states have more environmental enforcement power, that means that any differences in environmental policy efforts, such as those that we have observed as a result of combined policy mandates, are more consequential. If the states have more power, economic concerns have more influence, who controls state legislatures and the governor's office has more influence, and how powerful and capable state bureaucracies are has more influence. The details of environmental federalism that I describe in these chapters are critical details in understanding where we succeed and fail in facing environmental problems.

Additionally, this research continues to emphasize the importance of considering organizational structures and the effects of combining policy mandates, generally speaking. As I express throughout this book, efforts to combine policy mandates in the name of ending "wasteful redundancy," promoting improved communication between departments with related work, and spurring efficiency is not uncommon. However, the findings here provide additional support for Gilad's (2015) assertion that a combination "of multiple tasks and goals . . . requires [organizations] to prioritize some tasks over others" (593). Regardless of how complementary we may believe policy mandates like conservation, public health, and environmental enforcement to be, the simple combination of these tasks with one another likely leads to the prioritization of one set of tasks and programs over others. The interviews and content analyses in this book certainly reflect that proposition. These findings suggest that agency combinations should be undertaken carefully, with the acknowledgment that some tasks—particularly those that are less politically or socially popular—could suffer. Furthermore, even though it is beyond the scope of this book, what these combinations mean for conservation or public health programs is also of consequence. It is possible that these more politically attractive mandates, themselves, may be worse off when combined with another mandate, even if they receive more attention than environmental enforcement. What agency combinations mean for the execution of policy implementation continues to be an important question to consider for scholars of public policy and administration, broadly speaking.

Lastly, the combination of environmental protection with additional policy mandates is not isolated to the United States (Hopper 2019). For example, the French Ministry of Ecology, Energy, and Sustainable Development, and Spatial Planning and the German Federal Ministry for Environment, Nature Conservation and Nuclear Safety (BMU) are both examples of combinations that occur internationally. Do these combinations have a similar impact outside of the United States? Or, do the differences in federalism, bureaucratic and executive structure, and economic/social systems adjust the effects of environmental agency design? The American states offer a unique opportunity to explore how a number of different political, economic, and environmental contexts may affect how bureaucratic structures shape policy implementation. But, these findings are still limited by the restraints of considering these differences from a single-nation perspective. Continued research in this area should also center around the effects of combined environmental protection agencies from a comparative perspective.

In one of my earliest interviews with environmental agency workers, a woman who acted as her agency's liaison between environmental protection and an additional mandate expressed that she was surprised anyone outside the agency would be interested in how the combination could affect her jobs. She mentioned that prior to her work as liaison she had not been aware of how much of an impact the combination was having on how she saw her role as a regulator. However, her position as liaison made the effects much clearer, as she began the work of translating the values and norms of each part of the agency and to try and make the connections that really would make a combination worthwhile. "I still think it can work," she stated, "but I'm not convinced it's working right now." "We've lost touch with what we have in common and how our goals really overlap. It's ironic how we thought the combination would make things easier, but in some ways, it really just made us more aware of how different we are."

NOTES

1. The data for 2017 shows a sharp increase in penalty amounts to about $3 billion; however, this is somewhat misleading because the increase is due almost entirely to one fine—$2.8 billion—against Volkswagen due to emissions-rigging.

Appendix 1
Interview Methodology and Questionnaire

The interviews included in chapters 3 and 4 include quotes and paraphrased material taken from around twenty-seven hours of semi-structured interviews I performed with combined environmental agency workers between 2014 and 2019.

SAMPLE SELECTION

In order to select individuals for these interviews, I recorded email addresses from state bureaucratic directories for individuals working within combined environmental agencies between 2014–2019. Individuals solicited for interviews include workers from both sides of the agencies, including public health officials, conservation officials, and environmental protection officials. For each state's agency, the number of individuals coming from different departments was about equal. These individuals include inspectors, epidemiologists, engineers, and political appointees.

From those lists, I sent recruitment emails for which I garnered around a 10 percent response rate. For individuals who agreed to participate, I scheduled twenty-minute interview sessions either in person or by telephone. I am unable to provide more detailed information about exactly who was included in the interviews due to concerns about anonymity; however, it is important to note that the vast majority of individuals interviewed were not political appointees.

INTERVIEWS

As I note above, I scheduled in-person or telephone interviews for twenty-minute time slots. Some interviews were less than twenty minutes, due to personal preference of the interviewee. Others went on for nearly an hour.[1] In total, these conversations amounted to around twenty-seven hours of interview material.

Over the scheduled interview time, I asked a series of semi-structured questions for which I recorded answers by hand (typed). I did not use an audio recording during these interviews. That being said, any quotes included in the book were recorded as exact quotes and checked with interviewees for accuracy.

The list of questions I used for every interview are listed below.

INTERVIEW QUESTIONS SCRIPT

My name is JoyAnna Hopper, and I am an assistant professor in the Department of Politics at the University of the South. I am currently working on a research project, in which I explore how the combination of environmental enforcement with additional mandates (e.g., public health or natural resource conservation) affects the way agency workers see their tasks, goals, and priorities. As part of this research, I am conducting interviews with employees of environmental agencies with multiple mandates. Would you be willing to be interviewed for my project?

Your participation is voluntary, there are no foreseeable risks, and the interview will be used for academic research. The interview will take approximately no more than a half hour. With your permission, I will record our conversation and take written notes to document the interview.

Your participation would be greatly appreciated. Do you have any questions about the interview or research at this time?

1. Thank you for speaking with me. Could your please tell me your official job title?
2. How long have you currently been employed in that position?
3. What does a typical day consist of for you? For example, what tasks do you perform on a daily basis?
4. What task is most important in your everyday work?
5. What are the agency's goals?
6. Which goal do you think is the agency's number one priority?
7. What are your own personal work goals?

8. Which of your personal goals is your number one priority?
9. Do you feel as if your daily tasks are directly related to the agency's main goals?
   9a. (If yes . . .) Do you think this makes your job easier to perform? Why?
   9b. (If no . . .) Do you think this makes your job more difficult to perform? Why?
10. Do you think that each agency program receives a proportionate amount of resources or focus?
    10a. (If no) . . . which program receives the most resources/focus? The least?
11. Do you feel as if the agency is supported by the state legislature? What about the governor?
12. Do you feel as if the work *you* perform for the agency is supported by the state legislature? What about the governor?
13. (if applicable) Do you feel as if the combination of (natural resource conservation, energy, or public health) with environmental regulation affects what programs are focused on by your agency? In what way?
14. Do you prefer an agency in which environmental regulation is separate from (natural resource conservation, energy, or public health)? Why/why not?
15. Has there been any discussion regarding separating (natural resource conservation, energy, or public health) from environmental regulation or combining programs together? If so, could you describe that discussion for me?
16. Do you feel that you manage environmental protection with tools similar to the federal Environmental Protection Agency? If so, what tools are similar? If not, how does your agency differ?
17. Is there anything else you care to discuss with me?

NOTES

1. Interviewees were reminded at the twenty-minute mark that they were welcome to end the interview at this time. Interviews only continued upon consent of the interviewee.

Appendix 2

Summary Statistics and Data Sources for Analyses in Chapters 6–8

Variable	Mean	Std. Dv.	Min.	Max.	Data Source
Total Annual Penalty Amounts	703328.70	1854021	0	2940000000	EPA: ECHO
Number of Enforcement Actions per Number of Violations	2.64	6.61	0	84.80	EPA: ECHO
Combined Agency Design	0.40	0.49	0	1	National Association of Clean Air Agencies (NACAA)
Combined Natural Resource and Environmental Protection Agency Design (NREP)	0.30	0.46	0	1	National Association of Clean Air Agencies (NACAA)
Combined Public Health and Environmental Protection Agency Design (PHEP)	0.10	0.30	0	1	National Association of Clean Air Agencies (NACAA)
Unified Republican Control	0.43	0.49	0	1	Natl. Conference of State Legislatures
Number of Polluting Facilities	867.93	755.7183	63	3833	EPA: ECHO
Environmental Spending (in millions)	248.68	541.49	19.08	4941.39	State agency budgets & Environmental Council of the States (2017)
Population (in millions)	6.34	7.05	0.56	39.54	U.S. Census Bureau
State Square Area (logged)	-2.11	1.10	-5.52	0.54	U.S. Census Bureau
Air Quality	0.76	0.09	0.50	0.96	EPA: Air Quality Index Report
Bureaucratic Capacity	4.86	1.51	0	8	Pew: Grading the States Report (2008)
Change in GSP	1.27	2.02	-6.5	13.4	Bureau of Economic Analysis
Natural Resource Dependence	1.52	1.12	0	3	Hall and Kerr (1991)
Laws Limiting Agency Actions	0.38	0.49	0	1	Natl. Conference of State Legislatures
No Enforcement Primacy	0.12	0.33	0	1	Konisky and Woods (2012a)
Sierra Club Membership	2.21	1.01	0.80	3.8	Konisky and Woods (2012a)
Republican Control of Legislature	0.44	0.50	0	1	Natl. Conference of State Legislatures
Legislative Staff	677.76	628.92	86	2919	Natl. Conference of State Legislatures
Term Limits	0.30	0.46	0	1	Natl. Conference of State Legislatures
Seats in the Legislature	147.65	59.68	49	424	Natl. Conference of State Legislatures
Number of Poor Health Days	3.76	0.57	2.7	5.3	United Health Foundation

Appendix 3

Descriptions of Dataset and Independent Variables of Interest (Chapters 6–8)

DATASET DESCRIPTION

The data set I compiled for analyses in chapters 6–8 contains state-level environmental regulation statistics, including total dollar amounts of penalties assigned to polluters that are found to be in violation of regulatory standards and the number of pollution enforcement actions (informal or formal) taken by state environmental agencies per the number of violations discovered. In addition, the data set contains an air quality measure, a measure of bureaucratic capacity, and other state-level variables that I will explain in-depth throughout this section (see appendix B for the summary statistics and sources of each variable used in chapters 6, 7, and 8). The variables in the data set for the analyses in this chapter contain data collected from 2010–2017. Although the EPA provides some data that goes back further, compliance and enforcement factors are generally combined over five-year terms, making it difficult to determine penalty amounts per state, per year. Thus, I am restricted to using a shorter time period than may be optimal. That said, variables of interest, such as the partisan makeup of state institutions and economic conditions, do change over this time period, so it is sufficient for the purposes of this analysis.[1]

INDEPENDENT VARIABLES

Combined Environmental Agency Design

To test my expectation that the *combined agency design* should result in more collaborative and cooperative enforcement behavior with industry, I use two independent variables of note. The first is a simple dummy variable for whether or not a state's environmental protection programs are combined with an additional policy area (either public health *or* natural resource conservation) (Models 1 and 3). A zero represents the lack of that combination. The second independent variable breaks the combination variable into two dummy variables, one for the public health combination and one for the natural resource conservation combination (Models 2 and 4). The mini-EPA or pollution control agency design is the excluded category. As with the first independent variable, a zero represents the lack of these combination types.

To determine whether or not states combined these functions, I used the National Association of Clean Air Agencies' (NACAA) list of state-level contacts to determine which agencies dealt with ambient air monitoring at the state level. This specification is important, as I am using only air quality, monitoring, and enforcement statistics to measure state agency behavior.[2] I looked at all agency websites shown on NACAA's contact list to determine whether or not the kind of public health and natural resource conservation programs I listed in chapter 1 were included alongside traditional environmental regulation programs, such as air and water compliance. In all, as I show in chapter 1, five states combine public health and environmental regulation, and fifteen states combine natural resource conservation and environmental regulation.

Measures of Political Context

The measure of political context and the number of facilities measure in my analyses are important controls because they allow me to directly test my theoretical proposition that agency design should maintain an effect, even when outside pressures are considered. Although there are a number of ways to measure state political control, I use a dummy variable for unified Republican control (1) and the lack of unified Republican control (0). A one represents a state in which Republicans hold a majority in both houses of the legislature and hold the governorship. I use unified Republican control because it may be the case that governors' ability to oversee executive agencies is dependent on the oversight motivations of the state legislature and vice versa. Additionally, Hedge and Johnson (2002) find that unified Republican control has an effect on regulatory behavior, and it is the political control variable that exerted the strongest effect in my models.

State Economic Conditions

For a measure that accounts for the unique economic pressures faced by state-level bureaucrats, I use the number of polluting facilities, the state's annual change in gross product, and the states' economic dependence on natural resources. This should help to gauge the dependency of the state on polluting industries and resource conservation and help to control for economic pressures that may vary upon that dependence. Including the number of regulated facilities also helps to ensure that penalty amounts are considered within the context of how many regulated facilities are in the state, helping us to estimate per capita.

State Need for Environmental Regulation

In addition to these variables I include other variables that deal with state need for regulatory environmental programs and state capacity for implementing regulatory environmental programs. To measure state need, I include the population of the state and an air quality measure. To measure state capacity/ability, I include the physical size of the state, how much money is spent on environmental programs, a measure of bureaucratic capacity, the environmental ideology of the state (as measured by state Sierra Club membership—and also proxied by spending), the passage of laws limiting environmental state agency authority, and state primacy over environmental programs. I will elaborate briefly on the operationalization of state air quality and bureaucratic capacity, as there are a number of ways to measure these variables.

I measure state air quality by using the EPA's measure of "good air days." Good air days mean air pollution poses little to no risk. Specifically, I use the proportion of days labeled as good air days by the EPA to measure state air quality. I use this measure, as it incorporates a number of emissions considerations into a variable that is easy to interpret. To measure bureaucratic capacity/autonomy, I incorporate a suitable proxy that has been used in previous studies (see Nicholson-Crotty and Miller 2012, for example). This proxy is the overall management capacity grades of the states, as published in the 2008 "State Management Report Card" (Barrett and Greene 2008). All of the control variables included in my analyses are standard components of models of environmental agency behavior or are additions based on the findings of analyses in previous chapters.

Finally, I include lagged variables to account for how previous conditions may affect current enforcement decisions. First, I include lagged dependent variables for both sets of models, as it is important to consider that bureaucratic agencies will base their current penalty and enforcement assessments on the penalty amounts and enforcements issued previously. Additionally, I include a one-year lag for air quality. As I have mentioned in previous chapters, it is difficult to identify whether or not a decrease in enforcement actions is due to a preference for less enforcement or improving air quality. Thus, the lag for air quality helps control for the possibility that improving environmental conditions in the previous year are not dictating enforcement decisions in the current year.

NOTES

1. Robinson and Meier (2006) use a similar time span (four years) to observe path dependence. As they point out, four years of data is not enough to see the initial decision to "embark on one path or another," but it should be sufficient in showing the effects of different paths from one year to the next (248).

2. I use ambient air monitoring, rather than other areas of environmental regulation because of the consistency that exists in measurement, especially in comparison to the monitoring of water sources.

References

Ali, Safia Samee and Christina Wilkes. 2016. "Experts Say California's Environmental Policies are Bellwether for Economic Growth." *NBC News*.

American Public Health Association. 2016. "The State Public Health Agency." Online. https://www.apha.org/policies-and-advocacy/public-health-policy-state ments/policy-database/2014/07/22/10/14/the-state-public-health-agency. Accessed 20 November 2016.

"Ash Council Memo: Federal Organization for Environmental Protection." 1970. Executive Office of the President: President's Advisory Council on Executive Organization. Washington, D.C.

Association of State and Territorial Health Officials (ASTHO). 2016. "Programs." Online. http://www.astho.org/Programs.aspx. Accessed 20 November 2016.

Atlas, Mark. 2007. "Enforcement Principles and Environmental Agencies: Principal-Agent Relationships in a Delegated Environmental Program." *Law and Society Review* 41(4): 939–980.

Austin, John. 2017. "EPA Cuts Could Hurt Texas." *CNHI*. Online. HTML. https://www.cnhi.com/featured_stories/epa-cuts-could-hurt-texas/article_a4e30d3e-05b1 -11e7-8c45-0f1a527e9210.html. Accessed 1 March 2018.

Bacot, A. Hunter and Roy A. Dawes. 1997. "State Expenditures and Policy Outcomes in Environmental Program Management." *Policy Studies Journal* 25: 355–370.

Balkenbush, Andrea. 2014. "DNR 40 Years 1974-2014." *Missouri Resources* 31(3): 10–13.

Barker, Jacob. 2015. "Former DNR Employees Say Regulator Stifles Public Information." *St. Louis Post-Dispatch*. Online. http://www.stltoday.com/business/local /former-dnr-employees-say-regulator-stifles-public-information/article_6ff7ff31 -73c9-57b2-8986-ce9b945df032.html. Accessed 1 March 2016.

Barrett, Katherine and Richard Greene. 2008. "Grading the States '08: The Mandate to Measure." *The State Management Report Card for 2008*. Online. PDF. http://www.pewtrusts.org/~/media/legacy/uploadedfiles/pcs_assets/2008/Gradingthe States2008pdf.pdf.

Barrilleaux, Charles, Richard Feiock, and Robert E. Crew, Jr. 1992. "Measuring and Comparing American States' Administrative Characteristics." *State and Local Government Review* 24(1): 12–18.

Bengston, Shawn, Randy Blankinship, and Craig Bonds. 2003. "Texas Parks and Wildlife Department History, 1963–2003." Online. PDF. https://tpwd.texas.gov/publications/pwdpubs/media/pwd_rp_e0100_1144.pdf. Accessed 1 March 2018.

Bengtsson, Magnus, Yasuhiko Hotta, Shiko Hayashi, and Lewis Akenji. 2010. "The Four Main Types of Policy Instruments: Applications in Asia." Institute for Global Environmental Strategies.

Bennear, Lori Snyder. 2006. "Evaluating Management-Based Regulation: A Valuable Tool in the Regulatory Toolbox?" in *Leveraging the Private Sector: Management-based Strategies for Improving Environmental Performance*. Edited by Cary Coglianese and Jennifer Nash. Washington, D.C.: Resources for the Future Press.

Birner, Betty. 2012. "Does the Language I Speak Influence the Way I Think?" *Linguistic Society of America*. Online. http://www.linguisticsociety.org/content/does-language-i-speak-influence-way-i-think. Accessed 1 March 2016.

Burke, Thomas A., Nga L. Tran, and Nadia M. Sahlauta. 1995. "Identification of State Environmental Services: A Profile of the State Infrastructure for Environmental Health and Protection." John Hopkins University: School of Hygiene and Public Health, Department of Health Policy and Management, Division of Public Health.

Caiazzo, Fabio, Akshay Ashok, Ian A. Waitz, Steve H. L. Yim, and Steven R. H. Barrett. 2013. "Air Pollution and Early Deaths in the United States. Part I: Quantifying the Impact of Major Sectors in 2005." *Atmospheric Environment* 79: 198–208.

California Environmental Protection Agency. 2001. "The History of the CalEPA." Online. PDF. https://calepa.ca.gov/wp-content/uploads/sites/6/2016/10/About-History01-Report.pdf. Accessed 1 February 2019.

California Environmental Protection Agency. 2016. "Environmental Compliance and Enforcement Report." Online. PDF. https://calepa.ca.gov/wp-content/uploads/sites/6/2018/02/enforcement_report_2016.pdf. Accessed 1 February 2019.

California Environmental Protection Agency. 2017. "Environmental Compliance and Enforcement Report." Online. PDF. https://calepa.ca.gov/wp-content/uploads/sites/6/2019/02/enforcement_report_2017.pdf. Accessed 1 February 2019.

California Environmental Protection Agency. 2018. "Major Accomplishments: 2011–2018." Online. PDF. https://calepa.ca.gov/wp-content/uploads/sites/6/2019/03/CalEPA_Accomplishments_Report_2011-2018_a.pdf. Accessed 1 February 2019.

California Environmental Protection Agency, Department of Pesticide Regulation. 2018. "Evaluation of Chlorpyrifos as a Toxic Air Contaminant: Executive Summary." Online. PDF. https://www.cdpr.ca.gov/docs/whs/pdf/chlorpyrifos_exec_summary.pdf. Accessed 1 February 2019.

"California Green Innovation Index." 2017. 9th Ed. *Next Ten*. Online. PDF. https://www.next10.org/sites/default/files/2019-06/2017-CA-Green-Innovation-Index-2.pdf. Accessed 1 February 2019.

Carey, John M., Richard G. Niemi, and Lynda W. Powell. 2000. *Term Limits in the State Legislatures*. Ann Arbor: The University of Michigan Press.

Carey, John M., Richard G. Niemi, Lynda W. Powell, and Gary F. Moncrief. 2006. "The Effects of Term Limits on State Legislatures: A New Survey of the 50 States." *Legislative Studies Quarterly* 31(1): 105–134.

Carpenter, Daniel. 2001. *The Forging of Bureaucratic Autonomy: Reputations, Networks, and Policy Innovation in Executive Agencies: 1862–1928*. Princeton, NJ: Princeton University Press.

Carpenter, David. 2010. *Reputation and Power: Organizational Image and Pharmaceutical Regulation at the FDA*. Princeton: Princeton University Press.

Carrigan, Christopher. 2014. "Captured by Disaster? Reinterpreting Regulatory Behavior in the Shadow of the Gulf Oil Spill." In *Preventing Regulatory Capture: Special Interest Influence and How to Limit It*. Edited by Daniel Carpenter and David A. Moss. New York: Cambridge University Press.

Carrigan, Christopher. 2017. *Structured to Fail? Regulatory Performance under Competing Mandates*. Cambridge: Cambridge University Press.

Chaney, Carole Kennedy and Grace Hall Saltzstein. 1998. "Democratic Control and Bureaucratic Responsiveness: The Police and Domestic Violence." *American Journal of Political Science* 42(July): 745–768.

Chemical Watch. 2019. "New York State Delays Enforcement of Cleaning Products Disclosure: Industry Litigation, Concern with Programme Remains." Online. HTML. https://chemicalwatch.com/73669/new-york-state-delays-enforcement-of-cleaning-products-disclosure. Accessed 1 March 2019.

Chu, Jennifer. 2013. "Study: Air Pollution Causes 200,000 Early Deaths Each Year in the U.S." *MIT News*. Online. HTML. http://news.mit.edu/2013/study-air-pollution-causes-200000-early-deaths-each-year-in-the-us-0829. Accessed 1 August 2019.

Chun, Young Han and Hal G. Rainey. 2005. "Goal Ambiguity in U.S. Federal Agencies." *Journal of Public Administration Research and Theory* 15(1): 1–30

Coglianese, Cary and Robert A. Kagan. 2007. Introduction to *Regulation and Regulatory Processes*. Aldershot, UK: Ashgate Publishing.

Coglianese, Cary and David Lazer. 2003. "Management-based Regulation: Prescribing Private Management to Achieve Public Goals." *Law and Society Review* 37(4): 691–730.

Coglianese, Cary and Jennifer Nash. 2001. "Environmental Management Systems and the New Policy Agenda." In *Regulating from the Inside: Can Environmental Management Systems Achieve Policy Goals?* Edited by C. Coglianese and Jennifer Nash. Washington, D.C.: Resources for the Future Press.

Cohen, Dara Kay, Mariano-Florentino Cuellar, and Barry Weingast. 2006. "Crisis Bureaucracy: Homeland Security and the Political Design of Legal Mandates." *Stanford Law Review* 59(3): 673–760.

Collier, Kiah. 2017. "A Pass to Poison: How the State of Texas Allows Industrial Facilities to Repeatedly Spew Unauthorized Air Pollution—with Few Consequences." *The Texas Tribune*.

Collier, Kiah. 2019. "Why has Texas Suddenly Decided to Immediately Sue Industrial Polluters?" *The Texas Tribune*.

Colorado Air Quality Control Commission (Department of Public Health and Environment). 2017. "Report to the Public, 2017–2018." Online. PDF. https://drive

.google.com/file/d/1djnq35AqF3_vkOn5lJA9_BOuicwqNzyT/view. Accessed 1 March 2018.

Colorado Department of Public Health and Environment. 2015. "Healthy Colorado: Shaping a State of Health: Colorado's Plan for Improving Public Health and the Environment, 2015–2019." Online. PDF. https://www.colorado.gov/pacific/sites /default/files/OPP_2015-CO-State-Plan.pdf. Accessed 1 March 2018.

Colorado Department of Public Health and Environment. 2016. "Strategic Plan 2016– 2019 and Department Implementation Plan FY 2016–17." Online. PDF. https:// www.colorado.gov/pacific/sites/default/files/OPP_CDPHE-2016-2019-Strategic -Plan-FY2016-17.pdf. Accessed 1 March 2018.

Colorado Department of Public Health and Environment. 2019. "About the Solid and Hazardous Waste Commission." Online. HTML. https://www.colorado.gov /pacific/cdphe/about-solid-and-hazardous-waste-commission. Accessed 1 February 2019.

Colorado Department of Public Health and Environment. 2019. "Services and Information." Online. HTML. https://www.colorado.gov/pacific/cdphe/categories /services-and-information. Accessed 1 February 2019.

"Colorado's Public Health System . . . and a bit of History." *Colorado Association of Local Public Health Officials.* Online. PDF. http://www.calpho.org/up loads/6/8/7/2/68728279/co_public_health_history.pdf. Accessed 1 March 2018.

Daley, Dorothy and James C. Garand. 2005. "Horizontal Diffusion, Vertical Diffusion, and Internal Pressure in State Environmental Policymaking, 1989–1998." *American Politics Research* 37: 615–644.

Dam Concerned Citizens, Inc. 2009. "Is the Fox Guarding the Hen House? NYSDEC Fails to Attend Sept. 30, 2009 SPDES Meeting!" Online. HTML. https://users .midtel.net/dccinc/reports/spdes_meeting.html. Accessed 1 February 2018

Davies, J. Clarence, III. 1970. *The Politics of Pollution.* New York: Pegasus.

Davis, Angela K., Jeremy M. Piger, and Lisa M. Sedor. 2012. "Beyond the Numbers: Measuring the Information Content of Earnings Press Release Language." *Contemporary Accounting Research* 29(3): 845.

Davis, Charles and Sandra K. Davis. 1999. "State Enforcement of the Federal Hazardous Waste Program." *Polity* 31: 450–468.

Denhardt, Robert B. 2011. *Theories of Public Organization.* 6th ed. Boston, MA: Wadsworth.

Dewatripont, Mathias, Ian Jewitt, and Jean Tirole. 2000. "Multitask Agency Problems: Focus and Task Clustering." *European Economic Review* 44(4-6): 869–877.

Dillon, Karen. 2013. "Kansas Supreme Court Reverses Permit Granted to Sunflower Coal Plant." *The Kansas City Star.* Online. http://www.kansascity.com/news/local /article328852/Kansas-Supreme-Court-reverses-permit-granted-to-Sunflower -coal-plant.html. Accessed 1 March 2016.

Elling, Richard C. 1984. "State Legislative Influence in the Administrative Process: Consequences and Constraints." *Public Administration Quarterly* 7: 457–481.

Environmental Council of the States. 2010. "August 2010 Green Report." www.ecos .org/files/4157_file_August_2010_Green_Report.pdf.

Environmental Council of the States. 2017. "Status of State Environmental Agency Budgets (EAB), 2013–2015." Online. PDF. https://www.ecos.org/wp-content

/uploads/2017/03/Budget-Report-FINAL-3_15_17-Final-4.pdf. Accessed 1 March 2018.

Environmental Working Group. 2012. "Inside Track: Cuomo Team Gives Drillers Jump Start to Influence Fracking Rules." Online. HTML. https://www.ewg.org /research/inside-track. Accessed 1 Feburary 2018.

Fehling, Dave. 2014. "Texas Pollution Worsens as Budget Shrinks for Regulators." *NPR: State Impact.* Online. HTML. https://stateimpact.npr.org/texas/2014/04/22 /texas-pollution-worsens-as-budget-shrinks-for-regulators/. Accessed 1 March 2018.

Finley, Bruce. 2019. "Colorado Lets Oil and Gas Companies Pollute for 90 Days without Federally Required Permits that Limit Emissions." *The Denver Post.*

Garcia-Gonzales, D., S. B. C. Shonkoff, J. Hays, and M. Jerrett. 2019. "Hazardous Air Pollutants Associated with Upstream Oil and Natural Gas Development: A Critical Synthesis of Current Peer-Reviewed Literature." *Annual Review of Public Health* 40: 283–304.

Geuss, Megan. 2018. "EPA Says Auto Emissions Standards Are Too High, Questions California's Waiver." *ArsTechnica.* Online. HTML. https://arstechnica.com /cars/2018/04/epa-says-auto-emissions-standards-are-too-high-questions-califor nias-waiver/. Accessed 1 February 2019.

Gilad, S. 2015. "Political Pressures, Organizational Identity, and Attention to Tasks: Illustrations from Pre-Crisis Financial Regulation." *Public Administration* 93(3): 593–608.

Gilless, Keith, Robert Lee, Bruce Lippke, and Paul Sommers. 1990. "Three-State Impact of Spotted Owl Conservation and Other Timber Harvest Reductions: A Comparative Evaluation of the Economic and Social Impacts." Seattle: Institute of Forest Resources, College of Forest Resources, University of Washington. September (#69).

Glick, Daniel. 2019a. "Colorado's Vaunted Oil and Gas Rules Are Flawed and Inadequately Enforced." *Colorado Independent.*

Glick, Daniel. 2019b. "A Former Colorado Air Quality Inspector Speaks Out." *Colorado Independent.*

Goldie, S. 2003. "Chapter 15: Public Health Policy and Cost-Effectiveness Analysis." *JNCI Monographs* 2003(31): 102–110.

Gordon, L. J. 1991. "Reaching the Environmental Health Objectives." *Journal of Public Health Policy* 12(1): 5–9.

Gormley, William, John Hoadley, and Charles Williams. 1983. "Potential Responsiveness in the Bureaucracy: Views of Public Utility Regulation." *American Political Science Review* (September): 704–717.

Gregor, Katherine. 2010. "Environmental Cage Match: After a History of Pulling Its Punches, Is the EPA Finally Forcing TECQ to Clean Up the Texas Air?" *The Austin Chronicle.*

Grosse, Scott D., Steven M. Teutsch, and Anne C. Haddix. 2007. "Lessons from Cost-Effectiveness Research for United States Public Health Policy." *Annual Review of Public Health* 28: 365–391.

Haight, Alex. 2018. "NYSDEC Continues to Delay Report on Niagara Sanitation Site." *Spectrum News.* Online. HTML. https://spectrumlocalnews.com/nys/buf

falo/news/2018/12/12/nysdec-continues-to-delay-report-on-niagara-sanitation-site. Accessed 1 March 2019.

Hall, Bob and Mary Lee Kerr. 1991. *The 1991–1992 Green Index: A State-by-State Guide to the Nation's Environmental Health*. Island Press.

Hammond, Thomas H. and Jack H. Knott. 1996. "Who Controls the Bureaucracy? Presidential Power, Congressional Dominance, Legal Constraints, and Bureaucratic Autonomy in a Model of Multi-Institutional Policy-Making." *The Journal of Law, Economics, and Organization* 12(1): 119–166.

Harkins, J. F. and M. A. Baggs. 1987. "An Alternative to Public Health-Based Environmental Protection: A Comprehensive Environmental Protection Concept." *University of Kansas Law Review* 35(2): 431–441.

Harrington, W. and R. D. Morgenstern. 2007. "Economic Incentives Versus Command and Control: What's the Best approach for Solving Environmental Problems?" in *Acid in the Environment*. Edited by G. R. Visgilio and D. M. Whitelaw. Boston, MA: Springer.

Hays, Scott P., Michael Esler, and Carol F. Hays. 1996. "Environmental Commitment among the States: Integrating Alternative Approaches to State Environmental Policy." *Publius: The Journal of Federalism* 26: 41–58.

Hedge, David and Renee Johnson. 2002. "The Plot that Failed: The Republican Revolution and Political Control of the Bureaucracy." *Journal of Public Administration Research and Theory* 12(3): 333–351.

Henry, Terrence. 2013. "After West Fertilizer Explosion, Concerns Over Safety, Regulation and Zoning." *NPR: State Impact*. Online. HTML. https://stateimpact .npr.org/texas/2013/04/22/after-west-fertilizer-explosion-concerns-over-safety -regulation-and-zoning/. Accessed 1 March 2018.

Hooker, Brad and Sara Wyant. 2019. "California Regulators Ban Chlorpyrifos." *AgriPulse*. Online. HTML. https://www.agri-pulse.com/articles/12182-california -regulators-ban-chlorpyrifos. Accessed 1 June 2019.

Hopkins, Jamie Smith. 2015. "In Texas, Environmental Officials Align with Polluters." *National Geographic*.

Hopper, JoyAnna S. 2017. "The Regulation of Combination: The Implications of Combining Natural Resource Conservation and Environmental Protection." *State Politics and Policy Quarterly* 17(1): 105–124.

Hopper, JoyAnna S. 2019. "Having it All? The Implications of Public Health and Environmental Protection Partnerships in the American States." *Environmental Policy and Governance* 29(1): 35–45.

Hopper, JoyAnna Sutherlin. 2013. "The Environmental Health Paradox: How Combining Public Health with Environmental Protection May Tip the Balance in Favor of Public Health Programming." Master's Thesis. University of Missouri.

Hsiao, Cheng. 2003. *Analysis of Panel Data*. 2nd ed. Cambridge University Press.

Huber, John D., Charles R. Shipan, and Madelaine Pfahler. 2001. "Legislatures and Statutory Control of Bureaucracy." *American Journal of Political Science* 45(2): 330–345.

Huffman, James L. 2000. "The Past and Future of Environmental Law." *Environmental Law* 30(1): 23–33.

Humphries, M. 2004. *Oil and Gas Exploration and Development on Public Lands*. Washington, D.C.: Congressional Research Service.

Hunter, S. and Waterman R. 1992. "Determining an Agency's Regulatory Style: How does the EPA Water Office Enforce the Law?" *Western Political Quarterly* 45(2): 403–417.

Institute of Medicine. 1988. "A History of the Public Health System." In *Institute of Medicine Committee for the Study of the Future of Public Health*. Washington D.C.: National Academies Press.

Kansas Department of Health and Environment. 2013. "2013 Annual Report: Charting a Path for Quality Improvement in Public Health." Online. PDF. http://www.kdheks.gov/reports/2013_Annual_Report.pdf. Accessed 1 March 2016.

Kansas Department of Health and Environment. 2015. "KDHE 2015 Annual Report." Online. PDF. http://www.kdheks.gov/reports/Annual_Report_2015.pdf. Accessed 1 March 2016.

Kansas WRAPS: Watershed Restoration and Protection Strategy. 2011. "The Process." Online. http://www.kswraps.org/wraps-process. Accessed 1 March 2016.

Karkkainen, Bradley. 2001. "Information as Environmental Regulation: TRI, Performance Benchmarking, Precursors to a New Paradigm?" *Georgetown Law Journal* 89: 257–370.

Knickmeyer, Ellen. 2019. "EPA Enforcement Drops Sharply in Trump's 2nd Year in Office." *AP News*. Online. HTML. https://www.apnews.com/9d10456338af48dc918cbaa24ea6a4ce. Accessed 1 April 2019.

Konisky, David M. 2007. "Regulatory Competition and Environmental Enforcement: Is There a Race to the Bottom?" *American Journal of Political Science* 51(4): 853–872.

Konisky, David M. 2008. "Regulator Attitudes and the Environmental Race to the Bottom Argument." *Journal of Public Administration Research and Theory* 18(2): 321–344.

Konisky, David M. 2009. "The Limited Effects of Federal Environmental Justice Policy on State Enforcement." *Policy Studies Journal* 37(3): 475–496.

Konisky, David M. and Neal D. Woods. 2012a. "Environmental Policy." In *Politics in the American States: A Comparative Analysis*. 10th ed. Edited by Virginia Gray, Russell L. Hanson, and Thad Kousser. Thousand Oaks, CA: CQ Press.

Konisky, David M. and Neal D. Woods. 2012b. "Measuring State Environmental Policy." *Review of Policy Research* 29(4): 544–569.

Konisky, David M. and Neal D. Woods. 2018. "Environmental Federalism and the Trump Presidency: A Preliminary Assessment." *Publius: The Journal of Federalism* 48(3): 345–371.

Koontz, T. M. 2002. *Federalism in the Forest: National versus State Natural Resource Policy*. Washington D.C.: Georgetown University Press.

Kotchian, S. 1997. "Perspectives on the Place of Environmental Health and Protection in Public Health and Public Health Agencies. *Annual Review of Public Health* 18: 245–259.

Kousser, Thad and Justin H. Phillips. 2009. "Who Blinks First? Legislative Patience and Bargaining with Governors." *Legislative Studies Quarterly* 34: 55–86.

Kousser, Thad and Justin H. Phillips. 2010. "The Roots of Executive Power." Paper presented at the 2010 State Politics and Policy Conference, Springfield, Illinois.

Lang, Susan S. 2007. "Water, Air, and Soil Pollution Causes 40 percent of Deaths Worldwide, Cornell Research Survey Finds." *Cornell Chronicle.* Online. http://www.news.cornell.edu/stories/2007/08/pollution-causes-40-percent-deaths-worldwide-study-finds. Accessed 1 March 2016.

Laver, Michael, Kenneth Benoit, and John Garry. 2003. "Extracting Policy Positions from Political Texts Using Words as Data." *The American Political Science Review* 97(2): 311–331.

Layzer, Judith. 2015. *The Environmental Case: Translating Values into Policy.* 4th ed. Thousand Oaks, CA: CQ Press.

Lemov, Michael R. 1968. "Administrative Agency News Releases: Public Information versus Private Injury." *George Washington Law Review* 37: 63–81.

Lester, James P. 1995. *Environmental Politics and Policy: Theories and Evidence.* Durham, NC: Duke University Press.

Lewis, David. 2004. *President and the Politics of Agency Design.* Stanford, CA: Stanford University Press.

Lowry, William R. 1992. *The Dimensions of Federalism: State Governments and Pollution Control Policies.* Durham, NC: Duke University Press.

Luther, L. 2006. *The National Environmental Policy Act: Streamlining NEPA.* Washington, D.C.: Congressional Research Service.

Maclean, Alex. 2018. "State Water Plan Could Hurt Local Economy, Officials Say." *The Union Democrat.* Online. HTML. https://www.uniondemocrat.com/localnews/6401627-151/state-water-plan-could-hurt-local-economy-officials. Accessed 1 February 2019.

Martin, Phillip. 2013. "West Fertilizer Plant Explosion Leads to Scrutiny of Lack of State Regulation." *Progress Texas.* Online. HTML. https://progresstexas.org/blog/west-fertilizer-plant-explosion-leads-scrutiny-lack-state-regulation. Accessed 1 March 2018.

May, P. and S. Winter. 2000. "Reconsidering Styles of Regulatory Enforcement: Patterns in Danish Agro-Environmental Inspection." *Law and Policy* 22(2): 143–173.

Maynard-Moody, Steven and Adam W. Herbert. 1989. "Beyond Implementation: Developing an Institutional Theory of Administrative Policy Making." *Public Administration Review* 49(2): 137–143.

McCracken, Brewster. 2017. "Texas is a Surprising National Leader in Water Conservation." *The Cynthia and George Mitchell Foundation.* Online. HTML. https://cgmf.org/blog-entry/262/Texas-is-a-surprising-national-leader-in-water-conservation.html. Accessed 1 March 2018.

Meier, Kenneth J. and Laurence J. O'Toole. 2006. "Political Control versus Bureaucratic Values: Reframing the Debate." *Public Administration Review* 66(2): 177–192.

Metz, Christine. 2011. "Former KDHE Secretary Rod Bremby Says He Did Not Resign: Bremby Says 'Abuses' Occurred in Permit Process." *Lawrence Journal-World.*

Metzger, Luke. 2004. "Mandatory Fines Proven Clean Water Enforcement Tool." *Environment Texas*. Online. HTML. https://environmenttexas.org/news/txe/man datory-fines-proven-clean-water-enforcement-tool. Accessed 1 March 2018.

Miller, Daniel E. 2013. "Hudson River Estuary Habitat Restoration Plan." New York State Department of Environmental Conservation, Hudson River Estuary Program. Online. PDF. https://www.dec.ny.gov/docs/remediation_hudson_pdf/hrhrp.pdf. Accessed 1 February 2018.

Missouri Department of Natural Resources. 2012. "Collaborative Adaptive Management on Hinkson Creek." Online. PDF. http://helpthehinkson.org/documents/2012 -03-28HinksonCreekCAMfactsheet.pdf. Accessed 1 March 2016.

Missouri Department of Natural Resources. 2015. "2015-2020 Strategic Framework: Enhancing Missouri, It's in our Nature." Online. PDF. http://dnr.mo.gov/pubs /docs/strategicframework2015WEB.pdf. Accessed 1 March 2016.

Moe, Terry. 1989. "The Politics of Bureaucratic Structure." In *Can the Government Govern?* Edited by John E. Chubb and Paul E. Peterson. Brookings Institution.

Moncrief, Gary and Joel A. Thompson. 2001. "On the Outside Looking In: Lobbyists Perspectives on the Effects of State Legislative Term Limits." *State Politics and Policy Quarterly* 1: 394–411.

Moncrief, Gary and Peverill Squire. 2013. *Why States Matter: An Introduction to State Politics*. Rowman and Littlefield.

National Oceanic and Atmospheric Administration. 2019. "Climate Change Impacts." Online. HTML. https://www.noaa.gov/education/resource-collections/cli mate-education-resources/climate-change-impacts. Accessed 1 April 2019.

Nelson, Gabriel. 2010. "Kansas Approves Air Permit for Controversial Sunflower Coal-Fired Power Plant. *New York Times*.

New York Department of Environmental Conservation. 2015. "Final Supplemental Generic Environmental Impact Statement on the Oil, Gas and Solution Mining Regulatory Program." *Regulatory Program for Horizontal Drilling and High-Volume Hydraulic Fracturing to Develop the Marcellus Shale and Other Low-Permeability Gas Reservoirs*. Online. PDF. https://www.dec.ny.gov/docs/materi als_minerals_pdf/findingstatehvhf62015.pdf. Accessed 1 February 2018.

New York Department of Environmental Conservation. 2017. "Agency Highlights and Accomplishments for 2016 and 2017." Online. PDF. https://www.dec.ny.gov /docs/administration_pdf/decaccomplish2017.pdf. Accessed 1 February 2018.

New York Department of Environmental Conservation. 2018. "Agency Highlights and Accomplishments 2018." Online. PDF. https://www.dec.ny.gov/docs/adminis tration_pdf/decaccomplish2018.pdf. Accessed 1 March 2019.

New York Department of Environmental Conservation. "History of DEC." Online. HTML. https://www.dec.ny.gov/about/9677.html. Accessed 1 February 2018.

New York Department of Environmental Conservation. "History of DEC and Highlights of Environmental Milestones." Online. PDF. https://www.dec.ny.gov/docs /administration_pdf/dectimeline.pdf. Accessed 1 February 2018.

Nicholson-Crotty, Jill and Susan Miller. 2012. "Bureaucratic Effectiveness and Influence in the Legislature." *Journal of Public Administration Research and Theory* 22(2): 347–371.

Nickell, Stephen. 1981. "Biases in Dynamic Models with Fixed Effects." *Economet-rica* 49(6): 1417–1426.

Nie, Martin. 2008. "The Underappreciated Role of Regulatory Enforcement in Natural Resource Conservation." *Policy Sciences* 41(2): 139–164.

Norton, Bryan G. 2005. *Sustainability: A Philosophy of Adaptive Ecosystem Management*. Chicago, IL: The University of Chicago Press.

Off, Gavin. 2007. "Lessons in Leniency: In the 'Spirit of Compromise,' State Environmental Regulators Routinely Reduce Pollution Penalties Assessed against Large Animal Farms." *The Columbia Tribune*. Online. http://archive.columbiatribune.com/2007/may/20070506feat006.asp. Accessed 1 March 2016.

Ogden, Brian. 2013. "Pollution Plagues Hinkson Creek, but Tests Show the Water Is Improving." *Urban Pioneer: Missouri Urban Journalism Workshop*. Online. https://2013urbanpioneer.wordpress.com/2013/07/20/pollution-plagues-hinkson-creek-but-tests-show-the-water-is-improving/. Accessed 1 March 2016.

Omenn, G. S. 1996. "Putting Environmental Risks in a Public Health Context." *Public Health Reports (1974–)* 111(6): 514–516.

Parson, Edward A., Cary Coglianese, and Richard Zeckhauser. 2004. "Seeking Truth for Power: Informational Strategy and Regulatory Policymaking." *Minnesota Law Review* 2: 277–341.

Perry, J. L., A. M. Thompson, M. Tschirhart, D. Mesch, and G. Lee. 1999. Inside a Swiss Army Knife: An Assessment of AmeriCorps." *Journal of Public Administration Research and Theory* 9: 225–250.

Potoski, Matthew and Neal D. Woods. 2002. "Dimensions of State Environmental Policies: Air Pollution Regulation in the United States." *Policy Studies Journal* 30(2): 208–226.

Powell, Richard J. 2007. "Executive-Legislative Relations." In *Institutional Change in American Politics: The Case of Term Limits*. Edited by Karl Kurtz, Bruce Cain, and Richard Niemi. Ann Arbor: The University of Michigan Press, 134–147.

Rabe, Barry C. 2007. "Environmental Policy and the Bush Era: The Collision between the Administrative Presidency and State Experimentation." *Publius: The Journal of Federalism* 37: 413–431.

Riffe, Daniel, Stephen Lacy, and Frederick G. Fico. 2005. *Analyzing Media Messages Using Quantitative Content Analysis in Research*. Mahwah, NJ: Lawrence Erlbaum Associates.

Ringquist, Evan J. 1993. *Environmental Protection at the State Level*. Armonk, NY: M. E. Sharpe.

Ringquist, Evan J. 1995. "Is Effective Regulation Always Oxymoronic? The States and Ambient Air Quality." *Social Science Quarterly* 76: 69–87.

Robinson, Scott E. and Kenneth J. Meier. 2006. "Path Dependence and Organizational Behavior: Bureaucracy and Social Promotion. *The American Review of Public Administration* 36(3): 241–260.

Rosenthal, Alan. 1996. "State Legislative Development: Observations from Three Perspectives." *Legislative Studies Quarterly* 21: 161–198.

Rothman, Lily. 2017. "Here's Why the Environmental Protection Agency Was Created." *Time Magazine*.

Sadasivam, Naveena. 2018. "Dirtying the Waters: Texas Ranks First in Violating Water Pollution Rules." *Texas Observer*.

Sapat, Alka. 2004. "Devolution and Innovation: The Adoption of State Environmental Policy Innovations by Administrative Agencies." *Public Administration Review* 64: 141–151.

Sappington, David E. M. and Joseph E. Stiglitz. 1987. "Privatization, Information and Incentives." *Journal of Policy Analysis and Management* 6(4): 567–585.

Sarbaugh-Thompson, Marjorie, Lyke Thompson, Charles D. Elder, John Strate, and Richard C. Elling. 2004. *The Political and Institutional Effects of Term Limits*. New York: Palgrave.

Satchell, Michael. 1996. "At War in an Ancient Forest." *U.S. News and World Report*, 23 September, 74–76.

Satterlee, Lucas C. 2017. "Clearing the Fog: A Historical Analysis of Environmental and Energy Law in Colorado." *Villanova Environmental Law Journal* 28(1): 1–49.

Saunders, Richard E. 1937. "Centralizing Press Releases of the Government." *The Public Opinion Quarterly* 1(2): 101–103.

Schein, Edgar H. 1990. *Organizational Culture and Leadership*. 3rd ed. San Francisco, CA: Jossey-Bass; A Wiley Imprint.

Schmidt, Charles W. 2007. "Environment: California Out in Front." *Environmental Health Perspectives* 115(3): A144–A147.

Self-Walbrick, Sarah. 2018. "Stakeholders Oppose Sprinkler System Change that Could Prove Costly for Homeowners, Businesses." *Lubbock Avalanche-Journal*. Online. HTML. https://www.lubbockonline.com/news/local-news/busi ness/2018-01-24/stakeholders-oppose-sprinkler-system-change-could-prove -costly. Accessed 1 February 2019.

Shaw, Bryan W., Toby Baker, and Zak Covar. 2013. "TCEQ Penalty Policy: Effective April 1, 2014." TCEQ: Office of Compliance and Enforcement/Enforcement Division. Online. PDF. https://www.tceq.texas.gov/assets/public/comm_exec/pubs/rg /rg253/penaltypolicy2014.pdf.

Shepherd, Matthew D., George C. West, Tina S. Shoemaker, William L. Hargrove, and Robert F. St. Peter. 1999. *The Organization of Public Health and Environmental Functions in Kansas*. Topeka, KS: Kansas Health Institute.

Shover, N., D. A. Clelland, and J. Lynxwile. 1986. *Enforcement or Negotiation: Constructing a Regulatory Bureaucracy*. New York: SUNY Press.

Sigman, Hilary. 2003. "Letting States Do the Dirty Work: State Responsibility for Federal Environmental Regulation." *National Tax Journal* 56: 107–122.

Sinclair, Amber H. and Andrew B. Whitford. 2013. "Separation and Integration in Public Health: Evidence from Organizational Structure in the States." *Journal of Public Administration Research and Theory* 23(1): 55–77.

Smith, Sandy. 2016. "'Dangerously Close:' CSB Examines the Fatal Explosion in West, Texas." *EHS Today*. Online. HTML. https://www.ehstoday.com/emer gency-management/dangerously-close-csb-examines-fatal-explosion-west-texas . Accessed 1 March 2018.

Souder, Jon A. and Sally K. Fairfax. 1996. *State Trust Lands: History, Management, and Sustainable Use*. Lawrence: University Press of Kansas.

Sparks, Seth. 2015. "Are California's Emissions Regulations Too Strict?" *Trucks .com*. Online. HTML. https://www.trucks.com/2015/10/22/californias-emissions-regulations-strict/. Accessed 1 March 2018

Squire, Peverill. 1992. "Legislative Professionalism and Membership Diversity in State Legislatures." *Legislative Studies Quarterly* 17:69–79.

Squire, Peverill. 1998. "Membership Turnover and the Efficient Processing of Legislation." *Legislative Studies Quarterly* 23: 23–32.

Squire, Peverill. 2007. "Measuring State Legislative Professionalism: The Squire Index Revisited." *State Politics and Policy Quarterly* 7(2): 211–227.

Squire, Peverill. 2012. *The Evolution of American Legislatures: Colonies, Territories, and State, 1619–2009*. Ann Arbor, MI: University of Michigan Press.

Squire, Peverill. 2017. "A Squire Index Update." *State Politics and Policy Quarterly* 17: 361–371.

Staver, Anna. 2019. "Oil and Gas Generates $1 Billion in State and Local Taxes for Colorado, Report Finds: The Colorado Oil and Gas Association Report says the Industry Added about $13.5 Billion to Colorado's Domestic Product in 2017." *The Denver Post*.

Stewart, Jack. 2010. "Commentary: Regulation is Killing California Manufacturing." *The North Bay Business Journal*. Online. HTML. https://www.northbaybusiness journal.com/industrynews/4174435-181/commentary-regulation-is-killing-califor nia. Accessed 1 February 2019.

Stewart, Richard B. 2001. "A New Generation of Environmental Regulations?" *Capital University Law Review* 29(1): 21.

Tarantola, Andrew. 2018. "Regulation Has Helped, Not Hindered California's Green Economy." *Engadget*. Online. HTML. https://www.engadget.com/2018/05/16/reg ulation-has-helped-not-hindered-california-s-green-economy/. Accessed 1 February 2019.

Texas Commission on Environmental Quality. 2016. "Strategic Plan: Fiscal Years 2017–2021." Online. PDF. https://www.tceq.texas.gov/assets/public/comm_exec /pubs/sfr/035-17.pdf. Accessed 1 March 2018.

Texas Commission on Environmental Quality. 2018a. "Annual Report on Performance Measures: Fiscal Year 2018." Online. PDF. https://www.tceq.texas.gov /assets/public/comm_exec/pubs/sfr/055-18.pdf. Accessed 1 February 2019.

Texas Commission on Environmental Quality. 2018b. "Annual Enforcement Report: Fiscal Year 2018." Online. PDF. https://www.tceq.texas.gov/assets/public/compli ance/enforcement/enf_reports/AER/FY18/enfrptfy18.pdf. Accessed 1 February 2019.

"The Guardian: Origins of the EPA." 1992. *EPA Historical Publication-1*. https:// archive.epa.gov/epa/aboutepa/guardian-origins-epa.html.

Thomas, Stephanie. 2019. "Don't Look to Texas on Energy Deregulation." *The Hill*. Online. HTML. https://thehill.com/opinion/energy-environment/444866-dont -look-to-texas-on-energy-deregulation. Accessed 1 June 2019.

Thompson, Joel A. 1986. "State Legislative Reform: Another Look, One More Time, Again." *Polity* 18: 27–41.

Ting, Michael M. 2011. "Organizational Capacity." *Journal of Law, Economics, and Organization* 27(2): 245–271.

Toloken, Steve. 2019. "Federal Court Rules Formosa 'Serial Offender' in Plastic Pellet Pollution Case." *Plastics News*. Online. HTML. https://www.plasticsnews.com /news/federal-court-rules-formosa-serial-offender-plastic-pellet-pollution-case. Accessed 15 July 2019.

Tuma, Mary. 2017. "Anti-Choice Lawmakers Divert Millions from TCEQ to Crisis Pregnancy Centers." *Austin Chronicle*. HTML. https://www.austinchronicle. com/daily/news/2017-04-06/in-budget-debate-anti-choice-lawmakers-divert-mil lions-from-tceq-to-crisis-pregnancy-centers/. Accessed 1 March 2018.

United States Department of Health and Human Services. 2016. "Programs and Services." Online. HTML. https://www.hhs.gov/programs/index.html. Accessed 1 February 2016.

United States Department of Interior. 2019. "About." Online. HTML. https://www .doi.gov/whoweare. Accessed 1 February 2019.

United States Environmental Protection Agency. 2016a. "3.2 Command and Control." *National Center for Environmental Economics*. Online. HTML. https:// yosemite.epa.gov/ee/epa/eed.nsf/fa6512c6e51c4a208525766200639df2/9b6ed59f 910a89ea85257746000aff58!OpenDocument. Accessed 1 March 2016.

United States Environmental Protection Agency. 2016b. "Enforcement Basic Information." Online. HTML. https://www.epa.gov/enforcement/enforcement-basic -information. Accessed 1 March 2016.

United States Environmental Protection Agency. 2016c. "Enforcement Goals." Online. https://www.epa.gov/enforcement/enforcement-goals. HTML. Accessed 1 March 2016.

United States Environmental Protection Agency. 2016d. "Introduction: Environmental Enforcement and Compliance." Online. HTML. https://www3.epa.gov/region9 /enforcement/intro.html. Accessed 1 March 2016.

United States Environmental Protection Agency. 2018. "EPA's Audit Policy: Renewed Emphasis on Self-Disclosed Violation Policies." Online. HTML. https:// www.epa.gov/compliance/epas-audit-policy. Accessed 1 February 2019.

United States Environmental Protection Agency, Office of Inspector General (EPA/ OIG). 1998. *Consolidated Report on OECA's Oversight of Regional and State Air Enforcement Programs*. E1GAE7-03-0045-8100244. Washington, D.C.: EPA/ OIG.

United States Environmental Protection Agency, Office of Inspector General (EPA/ OIG). 2005. *Efforts to Manage Backlog Water Discharge Permits Need to be Accompanied Greater Program Integration*. Report No. 2005-P-00018. Washington, D.C.: EPA/OIG.

United States Environmental Protection Agency, Office of Inspector General (EPA/ OIG). 2011. *EPA Must Improve Oversight of State Enforcement*. Report No. 12-P-0113. Washington, D.C.: EPA/OIG.

United States Government Accounting Office (GAO). 2000. *Environmental Protection: More Consistency Needed among EPA Regions in Approach to Enforcement*. GAO/RCED-00-108. Washington, D.C.: GAO.

United States Government Accounting Office (GAO). 1991. *Testimony: Observations on EPA and State Enforcement under the Clean Water Act*. GAO/T-RCED-91-53. Testimony by Richard L. Hembra before the Subcommittee on Water Resources (U.S. House), May 14, 1991. Washington, D.C.: GAO.

United States Government Accounting Office (GAO). 2006. *Environmental Compliance and Enforcement: EPA's Effort to Improve and Make More Consistent its Compliance and Enforcement Activities*. GAO-06-840T. Testimony by John B. Stephenson before the U.S. Senate Committee on Environment and Public Works, June 28, 2006. Washington, D.C.: GAO.

U.S.D.A. Forest Service. 2002. "The Process Predicament: How Statutory, Regulatory, and Administrative Factors Affect National Forest Management." Washington, D.C.

Vogel, D. 1995. *Trading Up: Consumer and Environmental Regulation in a Global Economy*. London: Harvard University Press.

Vogel, D. 2018. *California Greenin': How the State Became an Environmental Leader*. Trenton, NJ: Princeton University Press.

Waterkeeper Alliance. 2017. "Environmental Groups Fight New York DEC's Toothless Permits for Industrial Animal Facilities." Online. HTML. https://waterkeeper.org/environmental-groups-fight-department-of-environmental-conservations-toothless-industrial-animal-facility-permits/. Accessed 1 February 2018.

Watertown Daily Times. 2016. "Full of Wind." Online. HTML. https://www.nny360.com/opinion/full-of-wind/article_f7f6b33f-2061-5897-bf20-b2997af818fa.html. Accessed 1 February 2018.

Wheeler, Andrew. 2019. "Ten Things You Should Know about Colorado's Oil and Gas Industry." *Conservation Colorado*. Online. HTML. https://conservationco.org/2019/02/28/blog-ten-things-you-should-know-cogcc/. Accessed 1 April 2019.

Wilson, James Q. 1989. *Bureaucracy: What Government Agencies Do and Why They Do It*. Basic Books.

Winslow, C. E. A. 1923. "The Evolution and Significance of the Modern Public Health Campaign." *Journal of Public Health Policy*. South Burlington, VT.

Wood, B. Dan. 1991. "Federalism and Policy Responsiveness: The Clean Air Case." *The Journal of Politics*. 53(3): 851–859.

Wood, B. Dan 1992. "Modeling Federal Implementation as a System: The Clean Air Case." *American Journal of Political Science* 36: 40–67.

Wood, Dan B. and John Bohte. 2004. "Political Transaction Costs and the Politics of Administrative Design." *Journal of Politics* 66(1): 176–202.

Woods, Neal D. 2006. "Primacy Implementation of Environmental Policy in the U.S. States." *Publius: The Journal of Federalism* 36: 259–276.

Woods, Neal D. 2008. "The Policy Consequences of Political Corruption: Evidence from State Environmental Programs." *Social Science Quarterly* 89: 258–271.

Woods, Neal D. and Michael Baranowski. 2006. "Legislative Professionalism and Influence on State Agencies: The Effects of Resources and Careerism." *Legislative Studies Quarterly* 31(4): 585–609.

Woods, Neal D., David M. Konisky, and Ann O'M Bowman. 2009. "You Get What You Pay For: Environmental Policy and Public Health." *Publius: The Journal of Federalism* 39: 95–116.

Wooldridge, J. M. 2009. *Introductory Econometrics: A Modern Approach.* 4th ed. Mason, OH: South-Western.

World Health Organization. 2014. "7 Million Premature Deaths Annually Linked to Air Pollution." Online. HTML. http://www.who.int/mediacentre/news/re leases/2014/air-pollution/en/. Accessed 1 March 2016.

Index

Abbott, Greg, 83
abortion, 84
accountability, 43, 79; bureaucracy and, 93
actions, xiii, 108, 146; agency values and, 111–25; compliance and, 72; goals and, 55; inaction and, 83; pressure and, 20–21; programs and, 105
agency culture, 89, 142; language and, 108
agency design, *9*, 102–4; enforcement and, 111–25, *122*; enforcement behavior and, 106; implications of, 11–14; measure of values and, 107–9; mini-EPAs and, xiv, 68, *137*; organizational capacity and, 127–42; PHEP and, 61–65; politics and, 107; results and variables in, 119–25, *120*
agency design choice, 7–10; employees and, 146; implications of, 11–14
agency values, 90–91; actions and, 111–25; agency design and, 107–9; measure of, 93–95
air pollution, 1, 81, 161; research on, 70; WHO on, 147
air quality, 105; states and, 106
allegations, 41, 113; leniency and, 34–35
American Canoe Association, 33

American Political Association, xv
approaches, 147; CDPHE and, 55–58; command and control approach, 69, 113; environmental agencies and, 89–109
a priori, 94–95
Ash, Roy L., 2
Ash memorandum (1970), 68
authority, 3, 83

Balkenbush, Andrea, 32
Baranowski, Michael, 129
Bennear, Lori Snyder, 147
best practices, 115
bias, 3
boards and commissions: states usage of, 68; Texas and control by, 8
Bremby, Rodney, 52
Brown, Edmund Brown Jr., 71
Brown, Jerry, 149
budget, 84; expenditures and, 15; KDHE and, 50
bureaucracy, 7, 89, 146; accountability and, 93; capacity of, 119, 131, 142n1; challenges for, 64; constraints on, 86; DOI programs and, 105; EPA and, 43; management capacity of, 123; relationships and, 21; reputation and, 131

bureaucratic workers: characteristics of, 90–91, 107; power and, 130

Bush Administration, 12

CAA. *See* Clean Air Act

CAFOs. *See* concentrated animal feeding operations

CalEPA. *See* California Environmental Protection Agency

California, 6; legislature in, 127; natural resources in, 70; statutes in, 73

California Environmental Protection Agency (CalEPA), 70–76, 138

California Green Innovation Index, 74

Carrigan, Christopher, 20

Carson, Rachel, 2

case study, 90, 121; CDPHE as, 53–54; documents and, 92; enforcement behavior and, 113; findings from, 124; KDHE as, 49–53; TCEQ as, 77–85, 137–38

CDPHE. *See* Colorado Department of Public Health and Environment

characteristics, 141; of bureaucratic workers, 90–91, 107; of states, 111

Chemical Watch, 40

chlorpyrifos, 75

citizens, 102; complaints by, 58–59; reflection of, 115

classification, 96

Clean Air Act (CAA), 147; compliance with, 53; interpretations of, 16; of 1970, 3

cleaning products, 40

Clean Water Act, 33

COGCC. *See* Colorado Oil and Gas Conservation Commission

collaboration, 63, 78; NYDEC and, 36–39

colleagues, xi

Colorado Department of Public Health and Environment (CDPHE), 48; as case study, 53–54

Colorado Oil and Gas Association, 58

Colorado Oil and Gas Conservation Commission (COGCC), 58; permits and, 59

Colorado Public Health Act (2008), 57

The Columbia Tribune, 35

combination: of activities, 10; agency design as, *122*; enforcement and, 29–45, 47–65; of policy, 12; regulation and, 150–51; variables and, 159–60

command and control approach, 69; EPA and, 113

commissions. *See* boards and commissions

communities, 58

complaints, 82; by citizens, 58–59

compliance, 21, 147; actions and, 72; assistance with, 63–64, 73; CAA and, 53; compliance assistance tools, 32–33; cost of, 69; data and, 124; as focus, 42; rates of, 80

concentrated animal feeding operations (CAFOs), 40

connectivity, 64

conservation: benefits of, 29; flexibility and legacy of, 31–34

context, 30, 64, 121; politics and, 86, 160

controversy: regulation and, 39–41; resource extraction and, 58–61

cooperation: flexibility and, 116; with industry, 50; NYDEC and, 36–39; partnerships and, 32

cost, 56; of compliance, 69; cost-effectiveness, 16, 37; to industry, 116

data, 125n1, 132, 142n1; analyses and, 116–19, *118*; collection of, 96–98; compliance and, 124; EPA and, 159; independent variables and, 159–62; penalization and, 151n1; states rank and, 22, *23–24*; statistics and, *158*

death: pollution and, 147; in United States, xii

About the Author

JoyAnna S. Hopper, PhD, is an assistant professor of politics at The University of the South in Sewanee, Tennessee. Her research and teaching interests include public policy and administration, subnational U.S. politics, and environmental enforcement behavior.

www.ingramcontent.com/pod-product-compliance
Lightning Source LLC
Chambersburg PA
CBHW022315280326
41932CB00010B/1103